Theory of the Hyperfine Structure of Free Atoms

Theory of the Hyperfine Structure of Free Atoms

LLOYD ARMSTRONG, Jr.

The Johns Hopkins University

WILEY-INTERSCIENCE
a Division of John Wiley & Sons, Inc.
New York • London • Sydney • Toronto

Copyright © 1971, by John Wiley & Sons, Inc.

All rights reserved. Published simultaneously in Canada.

No part of this book may be reproduced by any means, nor transmitted, nor translated into a machine language without the written permission of the publisher.

Library of Congress Catalog Card Number: 73-166316

ISBN 0-471-03335-9

Printed in the United States of America.

10 9 8 7 6 5 4 3 2 1

Preface

The student who wishes to learn the theory of atomic hyperfine structure finds himself facing several difficulties. First, a full understanding of this field requires knowledge of a large number of subjects, ranging from the relativistic theory of the atom, to Racah algebra, to many-body techniques. In addition, there is a lack of comprehensive reviews of the theory. The student is thus forced to learn by reading the original literature, where he may find that the many subfields of hyperfine structure theory are not always completely compatible. Authors from different specialities (or even within a speciality) may use different notation and make different approximations, which may not always be spelled out with great clarity. Thus the diversity of the field, which contributes greatly to its vitality, may make it very difficult for the student to combine in a consistent fashion the different aspects of the theory to obtain an overall view. The active worker in hyperfine structure is, of course, not immune to the difficulties produced by this diversity and may often be unwilling to spend the time required to sort out some aspect of the theory not directly concerned with his own research.

I have tried in this book to remove some of these barriers, first, by providing the background necessary for a study of the theory of atomic hyperfine structure and, second, by analyzing the entire field of interactions between the nucleus and electrons of an atom using a consistent and clearly stated set of approximations and a consistent set of techniques. The principal tools used in this exposition are elementary Lie theory and second-quantization techniques. These tools are chosen not only because they are particularly well suited to all of the subjects discussed, but also because their use allows one to see more clearly the relationships between the theory of hyperfine structure and the modern theory of atomic fine structure. These relationships are particularly important to understand, because the approximations made in carrying out fine structure calculations greatly affect the hyperfine structure calculations.

An attempt has been made to keep the material in this book sufficiently general for it to be accessible to an advanced graduate student, while at the same time providing enough detail to make it of value to the active worker in the field. I have endeavored to discuss in some depth most of the subjects that affect the hyperfine structure of free atoms. Obviously, not all subjects have been treated in equal detail; however, I have given numerous references to the original literature so that the interested reader himself can fill in the remaining gaps. Clearly, I have concerned myself only with the theory of the hyperfine structure of free atoms and have made no attempt to discuss such subjects as hyperfine interactions in solids and molecules and muonic hyperfine structure.

I thank Professor B. R. Judd for his encouragement and for his many helpful comments during the preparation of this manuscript.

LLOYD ARMSTRONG, JR.

Baltimore, Maryland
April 1971

Contents

I. **Classical Hyperfine Structure** — 1
   1. Introduction, 1
   2. Classical Hyperfine Interactions, 2

II. **Angular Momentum and Second Quantization** — 8
   1. Coupling of Two Angular Momenta, 8
   2. Coupling of Three Angular Momenta, 10
   3. Coupling of Four Angular Momenta, 13
   4. Spherical Tensors, 14
   5. Second-Quantization Techniques, 16
   6. Graphs, 18

III. **Atomic Structure** — 21
   1. Introduction, 21
   2. The Dirac Equation, 23
   3. Creation and Annihilation Operators for Relativistic Electrons, 27
   4. Groups and Group Generators, 30
   5. States, 34
   6. Perturbing Interactions, 40
   7. Operators and Representations, 45
   8. The Wigner-Eckart Theorem, 48
   9. Nonrelativistic Limit of the Perturbing Interactions, 49
   10. Discussion of Atomic Structure, 53

IV. **The Hyperfine Interaction** — 57
   1. Electromagnetic Fields of a Nonspherical Nucleus, 57
   2. The Hyperfine Hamiltonian, 62

3. The Hyperfine Hamiltonian for the Many-Electron Atom, 64
4. Effective Operators, 67
5. Interaction Constants and Nuclear Moments, 68
6. Nonrelativistic Limits of the Hyperfine Interactions, 73

V. The Central Field Approximation 79

1. Introduction, 79
2. The Nonrelativistic Central Potential, 79
3. The Relativistic Central Field, 84
4. Breakdown of the Central Field Approximation, 86
5. Perturbations in the Nonrelativistic Theory, 95
6. Core Polarization—Perturbation Analysis, 96
7. Core Polarization—Hartree-Fock Treatment, 107
8. Quadrupole Shielding, 110
9. Correlation, 116

VI. Hyperfine Structure in an External Field 122

1. External Magnetic Fields, 122
2. Diamagnetic Shielding, 129
3. External Uniform Electric Fields, 130
4. Interactions with the Gradient of an External Electric Field, 135

VII. Higher-Order Effects 137

1. Introduction, 137
2. Hyperfine Anomalies, 137
3. Breakdown of J as a Good Quantum Number, 145
4. Atomic Electric Dipole Moments, 150
5. Differential Polarizability, 156

VIII. Hyperfine Structure of the One-Electron Atom 160

1. Hyperfine Interaction Constants in One-Electron Atoms, 160
2. Relativistic Radial Integrals—Casimir Factors, 161
3. Approximate Hyperfine Interaction Constants, 165

IX. Hyperfine Structure in the Many-Electron Atom 167

1. Hyperfine Structure in the Configuration l^n, 167
2. The Half-Filled Shell, 174

3. The Contact Interaction in l^n, *175*
4. Hyperfine Structure in the Configuration $l^n l'^m$, *176*
5. Hyperfine Structure of Samarium, *179*

Appendix. Nonrelativistic Limits **189**

References **193**

Index **201**

Theory of the Hyperfine Structure of Free Atoms

I. Classical Hyperfine Structure

1. INTRODUCTION

Hyperfine structure has been known experimentally since the early days of spectroscopic studies of atoms. Michelson,[1] Fabry and Perot[2] and Lummer and Gehrcke[3] all found that many fine structure lines could be resolved in a series of very closely spaced lines; because this additional structure was so much smaller than the fine structure, it was called hyperfine structure. It was long felt that this new structure arose because mixtures of different isotopes were being used for spectroscopic studies; the assumption was made that each isotope emitted at a frequency slightly different from that of its neighbors because of changes in nuclear mass. This theory was not completely satisfactory because the magnitude of the measured hyperfine structure often appeared to be too great to be explained in this way. No real understanding of the effect was possible, however, so long as even the gross properties of atomic fine structure remained unexplained. Thus, it was not until 1924 that Pauli[4] suggested that hyperfine structure arose through interactions between the atomic electrons and the magnetic moment of the nucleus. Shortly after, Back and Goudsmit[5] succeeded in analyzing the hyperfine structure of bismuth in terms of coupling of angular momentum. In 1928, Jackson[6] calculated the magnetic moment of cesium from a study of its hyperfine structure. It perhaps helps to put these accomplishments into perspective if we recall that it was not until 1924 that de Broglie[7] suggested the wave nature of particles, and 1926 that Schrodinger[8] proposed his famous equation.

In the next few years, papers by Fermi,[9] Breit,[10] and Goudsmit[11] described theoretical studies of hyperfine structure using quantum mechanics and extended the studies to atoms with several electrons. Breit[12] and Racah[13] considered relativistic effects in hyperfine structure. Schmidt and Schuler[14] demonstrated the existence of an interaction between the electrons and a nuclear electric quadrupole moment; Casimir[15] studied these results and was able to calculate a value for the nuclear quadrupole moment. Casimir and Karreman[16] derived an expression for the interaction between a nuclear octupole moment and the electrons.

In the meantime, studies by Schuler and Keystone[17] and Kopfermann[18] had shown that there was a great deal of truth in the original assumptions concerning the nature of hyperfine structure: the fine structure splittings of different isotopes of the same element can indeed be slightly different. Thus in many cases two effects were superimposing a structure onto the fine structure. The effect caused by different isotopes having different fine structure energies has come to be called the isotope shift, with only the interaction between the nuclear moments and the electrons producing hyperfine structure.

In a series of papers in the 1940's, Racah[19-21] developed extremely elegant and powerful techniques for use in atomic calculations; Trees[22] applied these methods in 1953 to hyperfine structure calculations. Shortly thereafter, Schwartz[23] combined the Racah techniques with the Dirac equation to study relativistic hyperfine structure in one electron atoms. Extension of Schwartz's results to many electron atoms was extremely cumbersome until Sandars and Beck[24] introduced an effective operator approach in 1965.

Experimental techniques for measuring hyperfine interactions also have advanced tremendously over the years. The first real advance in technique over that of optical spectroscopy was the development by Rabi, et al. of atomic beam radiofrequency spectroscopy.[25,26] Using this method, very exact values can be obtained for the hyperfine constants in ground or low lying states of atoms. Hyperfine structure in some excited states can be measured by optical pumping and related techniques.[27] Hyperfine structure of atoms in solids and liquids can be studied in a variety of ways, using for example nuclear magnetic resonance,[28] paramagnetic resonance,[29] the Mossbauer effect,[30] and angular correlation.[31]

Atomic wavefunctions are usually obtained from a study of fine structure. Because hyperfine structure is so much smaller than fine structure, it can be considered as a perturbation on the fine structure levels. Use of the fine structure wavefunctions to calculate hyperfine structure effects is one of the most severe tests which can be applied to the wavefunctions. Thus, theoretical studies of hyperfine interactions are important not only because they enable one to explain observed structure, but also because they are a sensitive indicator of the accuracy of atomic wavefunctions.

2. CLASSICAL HYPERFINE INTERACTIONS

Derivations[32] of the fine structure interactions assume that the nucleus is a spherical point charge. As a result, the only interaction between the nucleus and the orbital electrons in an atom is of the form $-Ze^2/r$. In reality, the nucleus is neither a point nor spherical. It is instead a finite body which may deviate significantly from the spherical. The nucleus also possesses a spin I, which, as we shall show later, determines its allowed deformations. Hyperfine structure exists because of just this deviation of the nucleus from a spherical point charge. In the present section, we shall discuss simplified classical derivations of the main hyperfine interactions. It is hoped that this discussion will give the reader some feeling for the physical interactions which cause hyperfine structure.

The electrostatic energy of a nuclear charge distribution $\rho_n(r)$ placed in a potential field $\phi_e(r)$ produced by the atomic electron is

$$H_e = \int \rho_n(r_n) \phi_e(r_n) \, d\tau_n$$

where the electrons are assumed external to the nuclear volume. (We shall use Gaussian units in electromagnetic calculations.) The origin $(r = 0)$ shall always be taken to be the center of charge of the nucleus. Any other choice of origin might lead to nonphysical results, for example a nuclear electric dipole moment, which would have to be canceled out by higher-order perturbation theory. If $\phi_e(r)$ varies slowly over the nuclear volume, we can make the expansion

$$\phi_e(r_n) = \phi_e(0) + \mathbf{r}_n \cdot \nabla \phi_e(0) + \frac{1}{2} \sum_{i,j} x_i x_j \frac{\partial^2 \phi}{\partial x_i \, \partial x_j} + \cdots.$$

Using $\mathbf{E} = -\nabla \phi_e$ and $\nabla \cdot \mathbf{E} = 0$, this can be written

$$\phi_e(r_n) = \phi_e(0) - \mathbf{r}_n \cdot \mathbf{E}(0) - \frac{1}{6} \sum_{i,j} (3 x_i x_j - r^2 \delta_{ij}) \frac{\partial E_j(0)}{\partial x_i} + \cdots$$

or

$$H_e = Ze \phi_e(0) - \int \rho_n(r_n) \mathbf{r}_n \cdot \mathbf{E}(0) \, d\tau_n - \frac{1}{6} \sum_{i,j} Q_{ij} \frac{\partial E_j(0)}{\partial x_i} + \cdots \quad \text{(I-1)}$$

where Ze is the total nuclear charge, Q_{ij} is the quadrupole moment tensor

$$Q_{ij} = \int (3 x_i x_j - r^2 \delta_{ij}) \rho_n(r) \, d\tau.$$

The first term in H_e is the ordinary Coulomb interaction between a point nucleus and the electric charges. This is the only nonvanishing term in the expansion if the nucleus is assumed to be a spherical point charge. The second term vanishes because of parity considerations (see Section VI-6), leaving the third, the electric quadrupole interaction, as the largest perturbation to the fine structure. The quadrupole interaction can also be expressed as the dot product of two spherical tensors (see Eq. (II-18))

$$H_Q = \mathbf{Q}^{(2)} \cdot (\nabla \mathbf{E})^{(2)} \quad \text{(I-2)}$$

where

$$Q^{(2)}_{\pm 2} = \sqrt{\left(\frac{3}{8}\right)} \int \rho_n x_{n\pm}^2 \, d\tau_n$$

$$Q^{(2)}_{\pm 1} = \mp \sqrt{\left(\frac{3}{2}\right)} \int \rho_n z_n x_{n\pm} \, d\tau_n$$

$$Q^{(2)}_0 = \frac{1}{2} \int \rho_n (3z_n^2 - r_n^2) \, d\tau_n$$

which can be written as

$$\mathbf{Q}^{(2)} = \int \rho_n(r_n) r_n^2 \mathbf{C}^{(2)}(\theta_n, \phi_n) \, d\tau_n. \tag{I-3}$$

Here we have abbreviated $\mathbf{E}(0)$ as $\mathbf{E}$, and

$$(\mathbf{VE})^{(2)}_{\pm 2} = -\frac{\sqrt{6}}{12} \partial_\pm E_\pm$$

$$(\mathbf{VE})^{(2)}_{\pm 1} = \pm \frac{\sqrt{6}}{6} \partial_\pm E_z$$

$$(\mathbf{VE})^{(2)}_0 = -\frac{1}{2} \frac{\partial E_z}{\partial z}$$

with

$$x_\pm = x \pm iy$$

$$\partial_\pm = \frac{\partial}{\partial x} \pm i \frac{\partial}{\partial y}$$

$$E_\pm = E_x \pm i E_y.$$

The $C^{(k)}_q$ is $[4\pi/(2k+1)]^{1/2} Y_{kq}$, where Y_{kq} is the usual spherical harmonic.[33] The equivalence of these two forms can be shown by expanding Eq. (I-2) in cartesian coordinates and remembering that $\mathbf{V} \cdot \mathbf{E} = 0$.

To complete the calculation, we need only evaluate $(\mathbf{VE})^{(2)}$ in terms of the electronic charge density $\rho_e(r_e)$. The potential at the nucleus due to this charge is simply

$$\phi_e(r=0) = \int \frac{\rho_e(r_e)}{r_e} \, d\tau_e.$$

The electric field at the origin is of the form

$$E_z = -\frac{\partial \phi_e}{\partial z} = -\int \frac{\rho_e(r_e)}{r_e^3} z_e \, d\tau_e.$$

The gradient of the electric field is of the form

$$\nabla_z E_z = -\int \frac{\rho_e(r_e)}{r_e^5} (3z_e^2 - r_e^2)\, d\tau_e$$

or, in general

$$(\nabla \mathbf{E})_q^{(2)} = \int \frac{\rho_e(r_e)}{r_e^3} C_q^{(2)}(\theta_e, \phi_e)\, d\tau_e. \tag{I-4}$$

Moving electrons produce a magnetic field in accordance with the law of Biot and Savart.[34] We therefore should consider the interaction between this magnetic field (again assumed to arise from sources external to the nucleus) and the nucleus. Using elementary magnetostatics, we can express the force between the external field **B** and the nuclear current $\mathbf{J}_n$ as

$$\mathbf{F} = \frac{1}{c} \int \mathbf{J}_n(r_n) \times \mathbf{B}(r_n)\, d\tau_n$$

$$= -\frac{1}{c} \mathbf{B}(0) \times \int \mathbf{J}_n(r_n)\, d\tau_n + \frac{1}{c} \int \mathbf{J}(r_n) \times (\mathbf{r}_n \cdot \nabla)\mathbf{B}(0)\, d\tau_n + \cdots$$

where we have expanded the magnetic field around the origin. The first term on the right above vanishes for steady state currents ($\nabla \cdot \mathbf{J} = 0$), and the second can be rewritten utilizing

$$\mathbf{J} \times (\mathbf{r} \cdot \nabla)\mathbf{B} = -\nabla \times \mathbf{J}(\mathbf{r} \cdot \mathbf{B})$$

which is obtained using $\nabla \times \mathbf{B} = 0$ and remembering that ∇ operates only on **B**. Then, again using $\nabla \cdot \mathbf{J} = 0$ and the expansion of a triple vector product, we have

$$\mathbf{F} = \nabla \times (\mathbf{B}(0) \times \mathbf{m})$$

where **m**, the magnetic moment of the nucleus is defined as

$$\mathbf{m} = \frac{1}{2c} \int \mathbf{r}_n \times \mathbf{J}(r_n)\, d\tau_n. \tag{I-5}$$

Again using the expansion of a triple vector product, and $\nabla \times \mathbf{B} = \nabla \cdot \mathbf{B} = 0$ we find

$$\mathbf{F} = \nabla(\mathbf{m} \cdot \mathbf{B}(0)).$$

(The above calculation is described in greater detail in e.g. Jackson.[34]) The potential energy of a magnetic moment in an external field is therefore

$$H_m = -\mathbf{m} \cdot \mathbf{B}(0). \tag{I-6}$$

The magnetic field at the nucleus produced by the orbital motion of the electrons can be obtained using the Biot-Savart law:

$$\mathbf{B}_l = -\frac{e}{c}\sum_i \frac{\mathbf{v}_i \times (-\mathbf{r}_i)}{r_i^3} = -\frac{e\hbar}{mc}\sum_i \frac{\mathbf{l}_i}{r_i^3} = -2\mu_0 \sum_i \frac{\mathbf{l}_i}{r_i^3}$$

where l_i is the orbital angular momentum of the ith electron measured in units of $\hbar$, μ_0 is the Bohr magneton $e\hbar/2mc$. The electron itself behaves as if it were a spinning magnetic dipole with magnetic moment $-2\mu_0 \mathbf{s}$ (ignoring various quantum electrodynamical corrections). The factor of 2 enhancement of the magnetic moment over the usual expression which would pertain to a body with angular momentum $\mathbf{s}$ is a purely relativistic result. This result will be obtained in Sect. VI-1 from a study of the interaction between a relativistic electron and an external magnetic field. The spinning electron, because it possesses a magnetic moment, therefore produces a magnetic field at the nucleus of

$$\mathbf{B}_s = -\sum_i \frac{2\mu_0}{r_i^3}\left(3\frac{(\mathbf{r}_i \cdot \mathbf{s}_i)}{r_i^2}\mathbf{r}_i - \mathbf{s}_i\right).$$

The total field can then be written as

$$\mathbf{B} = \sum_i -\frac{2\mu_0}{r_i^3}\left(\mathbf{l}_i - \mathbf{s}_i + 3\frac{(\mathbf{r}_i \cdot \mathbf{s}_i)}{r_i^2}\mathbf{r}_i\right)$$

$$= \sum_i -\frac{2\mu_0}{r_i^3}(\mathbf{l}_i - \sqrt{10}(\mathbf{s}_i \mathbf{C}_i^{(2)})^{(1)}). \tag{I-7}$$

In the second step above, we have anticipated some of the results of Chapter II in order to express B in a more concise form.

We have, in the above derivation, assumed that all electrons are external to the nucleus. As we shall see in later chapters, this is not always true, especially for s electrons. (Angular momentum of electrons will often be given using spectroscopic notation, such that the letter s corresponds to $l = 0$; p, $l = 1$; d, $l = 2$; f, $l = 3$; etc.). There has been considerable controversy as to whether this s electron interaction should be considered a relativistic effect or not. It certainly is a relativistic effect insofar as p electrons can give rise to the same effect. In any case, we shall present here a classical derivation due to Ramsey[26] valid for the spherically symmetric density of an s electron; a more general derivation is given in Chapter IV. We first calculate the magnetic field inside a sphere with a radius only slightly larger than that of the nucleus. If we assume a uniform spin density throughout the volume of the sphere, there will be inside the sphere a uniform magnetization (magnetic moment per unit volume) of $\mathbf{M} = -2\mu_0|\psi_0|^2\mathbf{s}$ where ψ_0 is the value of the electronic wavefunction at $r = 0$. Elementary magnetostatics tells us that inside a sphere of uniform magnetization **M** is a field $\mathbf{B} = (8\pi/3)\mathbf{M}$. There will be no magnetic field at the nucleus due to the s electron density outside of the sphere because of the spherical symmetry of the electron density. We find then that an s electron produces a field

$$\mathbf{B}_C = -\frac{16\pi}{3}\mu_0|\psi_0|^2\mathbf{s} \tag{I-8}$$

at the nucleus; this value must be added to the field of Eq. (I-7) when s electrons are present. The interaction produced by $\mathbf{B}_C$ is often called the contact interaction, and is characterized by its dependence on the electronic wavefunction evaluated at the nuclear origin.

Higher-order terms also exist (see, e.g., Casimir[16]) but there is little reason to carry out a classical derivation as the magnetic dipole and electric quadrupole terms dominate hyperfine structure. Equations valid for interactions of any order will be obtained in Chapter IV using a relativistic derivation of the hyperfine equations.

The simple hyperfine interactions obtained above are often found insufficient to explain the very accurate experimental results which are available. It will therefore be our purpose in the remainder of this book to consider the many effects which can occur in the interaction between the electrons and the nucleus of a free atom. In addition to the contact interaction (Eq. (I-8)), there are several other effects in hyperfine structure which can be considered as relativistic in origin. Thus a thorough study of hyperfine structure must necessarily approach the problem from a relativistic standpoint. Relativistic calculations are, unfortunately, generally much more difficult to carry out than non-relativistic calculations. The complications arise because of both the form of the relativistic interactions and also the nature of the wavefunctions themselves. However, in Chapter III, we shall discuss techniques which make calculations of relativistic effects in complex atoms reasonably straightforward.

We shall not discuss nuclear models in this work, as that is a very broad subject indeed. We shall simply give the nuclear matrix elements which are to be evaluated; the reader interested in the subject of nuclear models will find many excellent works on the subject, for example Preston,[35] de-Shalit and Talmi,[36] and Bohr and Mottelson.[37]

II. Angular Momentum and Second Quantization

1. COUPLING OF TWO ANGULAR MOMENTA

The theory of angular momentum and spherical tensors will play a central role in our study of hyperfine structure. It is necessary, therefore, that we deviate somewhat from our main theme in order to briefly review some of the major aspects of this theory. Much more complete expositions can be found in for example Edmonds[33] and Brink and Satchler.[38]

From the standpoint of quantum mechanics, any set of three operators satisfying the commutation relations

$$[J_i, J_j] = J_i J_j - J_j J_i = i\varepsilon_{ijk} J_k \tag{II-1}$$

can be considered to be the components of an angular momentum vector. We define the angular momentum to be measured in units of $\hbar$; ε_{ijk} is $+1$ for an even permutation of the subscripts x, y, and z, -1 for an odd permutation. The most familiar example of angular momentum is the orbital angular momentum given classically by $\mathbf{L} = \mathbf{r} \times \mathbf{p}$, quantum mechanically (in units of $\hbar$) by the operator

$$L_i = -i\left(x_j \frac{\partial}{\partial x_k} - x_k \frac{\partial}{\partial x_j}\right).$$

Orthonormal eigenvectors can be obtained for any J; eigenvalues can easily be derived by considering the commutation relations (II-1). The eigenvalue equations are given by

$$J_z|J, M\rangle = M|J, M\rangle$$
$$\mathbf{J}^2|J, M\rangle = J(J + 1)|J, M\rangle$$
$$J_+|J, M\rangle = (J_x + iJ_y)|J, M\rangle = [J(J + 1) - M(M + 1)]^{1/2}|J, M + 1\rangle$$
$$J_-|J, M\rangle = (J_x - iJ_y)|J, M\rangle = [J(J + 1) - M(M - 1)]^{1/2}|J, M - 1\rangle$$
(II-2)

where arbitrary phases have been set in agreement with Condon and Shortley.[32] The allowed values of J and M are restricted to be either integer or half-integer numbers. M is often called the magnetic quantum number because the strength of the interaction of an atom with an external magnetic field is proportional to this number (see Section (VI-1)).

Wavefunctions which are eigenfunctions of angular momentum operators can be coupled together using 3-j symbols:

$$|j_1 m_1\rangle|j_2 m_2\rangle = \sum_{j,m}(-1)^{j_2-j_1-m}[j]^{1/2}\begin{pmatrix} j_1 & j_2 & j \\ m_1 & m_2 & -m \end{pmatrix}|j_1 j_2 jm\rangle \quad \text{(II-3a)}$$

where the quantity in parenthesis is a 3-j symbol. Terms written as $[a, b, \ldots]$ imply $(2a + 1)(2b + 1)\ldots$. The wavefunction $|j_1 j_2 jm\rangle$ is an eigenfunction of the angular momentum operator $\mathbf{j} = \mathbf{j}_1 + \mathbf{j}_2$ with eigenvalues

$$\mathbf{j}^2|j_1 j_2 jm\rangle = j(j+1)|j_1 j_2 jm\rangle$$
$$j_z|j_1 j_2 jm\rangle = m|j_1 j_2 jm\rangle.$$

The magnitude of j is limited by the usual rules of vector addition to the range $|j_1 - j_2| \leq j \leq j_1 + j_2$ with the sum $j_1 + j_2 + j$ an integer. These conditions limiting the possible values of j are known as the triangular conditions, indicating that one must be able to form a triangle out of the vectors $\mathbf{j}_1, \mathbf{j}_2$, and $\mathbf{j}$. If the transformation (II-3a) from the coupled to the uncoupled states is to be unitary, we must also have

$$|j_1 j_2 jm\rangle = \sum_{m_1, m_2}[j]^{1/2}(-1)^{j_2-j_1-m}\begin{pmatrix} j_1 & j_2 & j \\ m_1 & m_2 & -m \end{pmatrix}|j_1 m_1\rangle|j_2 m_2\rangle. \quad \text{(II-3b)}$$

The value of the 3-j symbol can be expressed as a general function of the arguments (see e.g. Edmonds[33]). In addition, extensive tables of numerical values of 3-j symbols have been given by Rotenberg et al.,[39] and algebraic expressions valid for certain values of the arguments have been given by Edmonds.[33] The 3-j symbol is related to the Clebsch-Gordan coefficient by

$$\begin{pmatrix} j_1 & j_2 & j_3 \\ m_1 & m_2 & m_3 \end{pmatrix} = (-1)^{j_1-j_2-m_3}[j_3]^{-1/2}(j_1 m_1 j_2 m_2|j_1 j_2 j_3 - m_3).$$

The 3-j symbol exhibits many symmetry properties. An even permutation of the columns does not change the numerical value of the symbol, whereas an odd permutation of columns merely introduces a phase factor equal to the sum of the coefficients in the top row of the symbol. Thus

$$\begin{pmatrix} j_1 & j_2 & j_3 \\ m_1 & m_2 & m_3 \end{pmatrix} = \begin{pmatrix} j_2 & j_3 & j_1 \\ m_2 & m_3 & m_1 \end{pmatrix}$$

but

$$\begin{pmatrix} j_1 & j_2 & j_3 \\ m_1 & m_2 & m_3 \end{pmatrix} = (-1)^{j_1+j_2+j_3} \begin{pmatrix} j_1 & j_3 & j_2 \\ m_1 & m_3 & m_2 \end{pmatrix}.$$

The coefficients in the lower row of the symbol must sum to zero

$$m_1 + m_2 + m_3 = 0$$

else the symbol vanishes. The signs of these coefficients in the lower row can be changed with a resulting change in the phase of the symbol, that is

$$\begin{pmatrix} j_1 & j_2 & j_3 \\ m_1 & m_2 & m_3 \end{pmatrix} = (-1)^{j_1+j_2+j_3} \begin{pmatrix} j_1 & j_2 & j_3 \\ -m_1 & -m_2 & -m_3 \end{pmatrix}.$$

Finally, because eigenvectors of the angular momentum operators which have different eigenvalues are orthogonal, we can obtain the conditions

$$\sum_{m_1, m_2} \begin{pmatrix} j_1 & j_2 & j \\ m_1 & m_2 & m \end{pmatrix} \begin{pmatrix} j_1 & j_2 & j' \\ m_1 & m_2 & m' \end{pmatrix} = \frac{\delta(j,j')\,\delta(m,m')}{[j]}$$

and (II-4)

$$\sum_{j, m} [j] \begin{pmatrix} j_1 & j_2 & j \\ m_1 & m_2 & m \end{pmatrix} \begin{pmatrix} j_1 & j_2 & j \\ m'_1 & m'_2 & m \end{pmatrix} = \delta(m_1, m'_1)\,\delta(m_2, m'_2)$$

from Eqs. (II-3). Several useful relations concerning 3-j symbols are given by Brink and Satchler.[38]

2. COUPLING OF THREE ANGULAR MOMENTA

The coupling of three angular momenta j_1, j_2, and j_3 can proceed in two distinctly different ways. The momenta j_1 and j_2 can first be coupled to a value J_{12}, with a subsequent coupling of J_{12} to j_3 to form a final J. Alternatively, j_2 and j_3 can be coupled to J_{23}, followed by a coupling of j_1 and J_{23} to the final J. In either case, the final state can be expanded in terms of the one particle states $|j_1 m_1\rangle$, $|j_2 m_2\rangle$ and $|j_3 m_3\rangle$ by use of Eqs. (II-3). Thus, for example

$$|j_1,(j_2j_3)J_{23},JM\rangle = \sum_{m_1,M_{23}}(-1)^{J_{23}-j_1-M}\begin{pmatrix}j_1 & J_{23} & J\\ m_1 & M_{23} & -M\end{pmatrix}[J]^{1/2}$$

$$\times |j_1m_1\rangle|(j_2j_3)J_{23}M_{23}\rangle$$

$$= \sum_{\substack{m_1,m_2\\M_{23},m_3}}(-1)^{J_{23}-j_1-M+j_3-j_2-M_{23}}$$

$$\times [J,J_{23}]^{1/2}\begin{pmatrix}j_1 & J_{23} & J\\ m_1 & M_{23} & -M\end{pmatrix}$$

$$\times \begin{pmatrix}j_2 & j_3 & J_{23}\\ m_2 & m_3 & -M_{23}\end{pmatrix}|j_1m_1\rangle|j_2m_2\rangle|j_3m_3\rangle.$$

An equivalent expansion for $|(j_1j_2)J_{12},j_3,JM\rangle$ reveals that the two methods of coupling three angular momenta can be related by the equation

$$|j_1,(j_2j_3)J_{23},JM\rangle = \sum_{J_{12}}(-1)^{j_1+j_2+j_3+J}\begin{Bmatrix}j_3 & J & J_{12}\\ j_1 & j_2 & J_{23}\end{Bmatrix}[J_{12},J_{23}]^{1/2}$$

$$\times |(j_1j_2)J_{12},j_3,JM\rangle \quad \text{(II-5)}$$

where

$$\begin{Bmatrix}j_1 & j_2 & j_3\\ m_1 & m_2 & m_3\end{Bmatrix}\begin{Bmatrix}j_1 & j_2 & j_3\\ l_1 & l_2 & l_3\end{Bmatrix} = \sum_{n_1,n_2,n_3}(-1)^{l_1+l_2+l_3+n_1+n_2+n_3}$$

$$\times \begin{pmatrix}j_1 & l_2 & l_3\\ m_1 & n_2 & -n_3\end{pmatrix}\begin{pmatrix}l_1 & j_2 & l_3\\ -n_1 & m_2 & n_3\end{pmatrix}\begin{pmatrix}l_1 & l_2 & j_3\\ n_1 & -n_2 & m_3\end{pmatrix}. \quad \text{(II-6)}$$

A number of useful equations can be obtained from Eq. (II-6) by multiplying both sides by various 3-j symbols and summing. Thus, multiplying both sides by

$$\sum_{m_1,m_2}\begin{pmatrix}j_1 & j_2 & j_3'\\ m_1 & m_2 & m_3'\end{pmatrix}$$

we find

$$\begin{Bmatrix}j_1 & j_2 & j_3\\ l_1 & l_2 & l_3\end{Bmatrix} = \sum_{\substack{\text{all }n\\m_1,m_2}}(-1)^{l_1+l_2+l_3+n_1+n_2+n_3}[j_3]\begin{pmatrix}j_1 & j_2 & j_3\\ m_1 & m_2 & m_3\end{pmatrix}$$

$$\times \begin{pmatrix}j_1 & l_2 & l_3\\ m_1 & n_2 & -n_3\end{pmatrix}\begin{pmatrix}l_1 & j_2 & l_3\\ -n_1 & m_2 & n_3\end{pmatrix}\begin{pmatrix}l_1 & l_2 & j_3\\ n_1 & -n_2 & m_3\end{pmatrix}. \quad \text{(II-7)}$$

The symbol on the left of Eq. (II-7) is known as a 6-j symbol; it is related to the W symbol of Racah[19] by

$$\begin{Bmatrix}j_1 & j_2 & j_3\\ l_1 & l_2 & l_3\end{Bmatrix} = (-1)^{j_1+j_2+l_1+l_2}W(j_1j_2l_2l_1;j_3l_3).$$

Because the 3-*j* symbols which define the 6-*j* symbol display such symmetry, one would expect the 6-*j* symbol itself to display a great deal of symmetry. In fact, the 6-*j* symbol is left unaffected by interchange of any two rows, or of upper and lower arguments in each of any two rows. Thus

$$\begin{Bmatrix} j_1 & j_2 & j_3 \\ l_1 & l_2 & l_3 \end{Bmatrix} = \begin{Bmatrix} j_2 & j_1 & j_3 \\ l_2 & l_1 & l_3 \end{Bmatrix} = \begin{Bmatrix} l_1 & j_2 & l_3 \\ j_1 & l_2 & j_3 \end{Bmatrix}.$$

In addition, the triangular conditions of the 3-*j* symbols impose triangular conditions on the 6-*j* symbols. These may be indicated by the following graphs:

Tables of numerical values of 6-*j* symbols have been given by Rotenberg, et al.[39] Algebraic formulae valid for special cases have been given by Edmonds.[33] A particularly important special case occurs when one of the arguments is zero:

$$\begin{Bmatrix} j_1 & j_2 & j_3 \\ l_1 & l_2 & 0 \end{Bmatrix} = \frac{\delta(j_1, l_2)\,\delta(j_2, l_1)}{[j_1, j_2]^{1/2}} (-1)^{j_1 + j_2 + j_3}. \qquad \text{(II-8)}$$

The orthogonality of different states leads to a sum rule on 6-*j* symbols

$$\sum_x [x] \begin{Bmatrix} a & b & x \\ c & d & p \end{Bmatrix} \begin{Bmatrix} a & b & x \\ c & d & q \end{Bmatrix} = \frac{\delta(p, q)}{[p]}. \qquad \text{(II-9)}$$

The reader may have noticed a third possible coupling of the three angular momenta; this third coupling is given by the scheme

$$\mathbf{j}_1 + \mathbf{j}_3 = \mathbf{J}_{13}$$
$$\mathbf{J}_{13} + \mathbf{j}_2 = \mathbf{J}.$$

Comparison of this scheme with the previous two coupling schemes leads immediately to another sum rule

$$\sum_x (-1)^{p+q+x} [x] \begin{Bmatrix} a & b & x \\ c & d & p \end{Bmatrix} \begin{Bmatrix} a & b & x \\ d & c & q \end{Bmatrix} = \begin{Bmatrix} c & a & q \\ d & b & p \end{Bmatrix}. \qquad \text{(II-10)}$$

Other sums can be evaluated by letting $q = 0$ in Eqs. (II-9) and (II-10), leading to

$$\sum_x [x] \begin{Bmatrix} a & b & x \\ b & a & p \end{Bmatrix} (-1)^x = \delta(p, 0)(-1)^{a+b} [a, b]^{1/2} \qquad \text{(II-11)}$$

$$\sum_x [x] \begin{Bmatrix} a & b & x \\ a & b & p \end{Bmatrix} = (-1)^{2(a+b)}.$$

Numerous other sum rules have been given by for example Judd,[40] Brink and Satchler,[38] and Yutsis, et al.[41] A particularly useful sum rule has been given by Biedenharn and Elliott[42]

$$\sum_x [x](-1)^{x+p+q+r+a+b+c+d+e+f} \begin{Bmatrix} a & b & x \\ c & d & p \end{Bmatrix} \begin{Bmatrix} c & d & x \\ e & f & q \end{Bmatrix}$$
$$\times \begin{Bmatrix} e & f & x \\ b & a & r \end{Bmatrix} = \begin{Bmatrix} p & q & r \\ e & a & d \end{Bmatrix} \begin{Bmatrix} p & q & r \\ f & b & c \end{Bmatrix}. \quad \text{(II-12)}$$

3. COUPLING OF FOUR ANGULAR MOMENTA

The coupling of four angular momenta, like the coupling of three, can proceed in a number of different ways. As in the case of three angular momenta, the coefficients describing the transformations between these different couplings can be found by uncoupling the states and comparing the results for different coupling schemes. We are interested here in coupling schemes in which pairs of angular momenta are coupled together, with the resultant angular momenta then coupled to the final value. This type of coefficient is important because for example it describes the transformation from LS to jj coupling—that is, a transformation from a state $|(s_1s_2)S, (l_1l_2)L, JM\rangle$ to a state $|(s_1l_1)j_1, (s_2l_2)j_2, JM\rangle$. Such a transformation is given in general by the equation

$$|(j_1j_2)J_{12},(j_3j_4)J_{34}, JM\rangle = \sum_{J_{13},J_{24}} [J_{12},J_{34},J_{13},J_{24}]^{1/2}$$
$$\times \begin{Bmatrix} j_1 & j_2 & J_{12} \\ j_3 & j_4 & J_{34} \\ J_{13} & J_{24} & J \end{Bmatrix} |(j_1j_3)J_{13},(j_2j_4)J_{24}, JM\rangle. \quad \text{(II-13)}$$

The quantity in curly brackets above is called a 9-j symbol; it is most concisely defined in terms of 6-j symbols as

$$\begin{Bmatrix} j_{11} & j_{12} & j_{13} \\ j_{21} & j_{22} & j_{23} \\ j_{31} & j_{32} & j_{33} \end{Bmatrix} = \sum_x (-1)^{2x}[x] \begin{Bmatrix} j_{11} & j_{21} & j_{31} \\ j_{32} & j_{33} & x \end{Bmatrix}$$
$$\times \begin{Bmatrix} j_{12} & j_{22} & j_{32} \\ j_{21} & x & j_{23} \end{Bmatrix} \begin{Bmatrix} j_{13} & j_{23} & j_{33} \\ x & j_{11} & j_{12} \end{Bmatrix}. \quad \text{(II-14)}$$

Each row and column of the 9-j symbol must satisfy the triangular conditions. An even permutation of rows or columns leaves the value of the 9-j symbol unchanged, as does a reflection about a diagonal. An odd permutation introduces a phase equal to -1 raised to a power equal to the sum of all of the arguments of the symbol.

An important simplification results when one argument of the 9-j symbol is zero:

$$\begin{Bmatrix} j_{11} & j_{12} & j_{13} \\ j_{21} & j_{22} & j_{23} \\ j_{31} & j_{32} & 0 \end{Bmatrix} = (-1)^{j_{12}+j_{21}+j_{13}+j_{31}} \frac{\delta(j_{13},j_{23})\,\delta(j_{31},j_{32})}{[j_{13},j_{31}]^{1/2}}$$

$$\times \begin{Bmatrix} j_{11} & j_{12} & j_{13} \\ j_{22} & j_{21} & j_{31} \end{Bmatrix}. \qquad \text{(II-15)}$$

Orthonormality of states leads to the sum rule

$$\sum_{x,y} [x,y] \begin{Bmatrix} a & b & x \\ c & d & y \\ e & f & g \end{Bmatrix} \begin{Bmatrix} a & b & x \\ c & d & y \\ e' & f' & g \end{Bmatrix} = \frac{\delta(e,e')\,\delta(f,f')}{[e,f]}. \qquad \text{(II-16)}$$

Numerous other relationships involving 9-j symbols are given by Judd,[40] Brink and Satchler,[38] and Yutsis, et al.[41] Descriptions of higher n-j coefficients are also given by Yutsis, et al., along with techniques for carrying out involved sums containing various n-j symbols.

4. SPHERICAL TENSORS

Racah[19] defined a spherical tensor of rank k as an operator having the same commutation relations with respect to the angular momentum operators as does the spherical harmonic of rank k. Thus the $2k+1$ components of a spherical tensor of rank k in J space ($\mathbf{T}^{(k)}$) satisfy the equations

$$[J_z, T_q^{(k)}] = q T_q^{(k)}$$
$$[J_\pm, T_q^{(k)}] = [k(k+1) - q(q\pm 1)]^{1/2} T_{q\pm 1}^{(k)}. \qquad \text{(II-17)}$$

Let us consider two tensor operators acting in J space, $\mathbf{T}^{(k)}$ and $\mathbf{U}^{(k')}$. We construct the combination

$$N_Q^{(K)} = \sum_{qq'} T_q^{(k)} U_{q'}^{(k')} \begin{pmatrix} k & k' & K \\ q & q' & -Q \end{pmatrix} (-1)^{k'-k-Q} [K]^{1/2}$$

and consider the commutators of N^K with $\mathbf{J}$. We find that

$$[J_z, N_Q^{(K)}] = Q N_Q^{(K)}$$
$$[J_\pm, N_Q^{(K)}] = [K(K+1) - Q(Q\pm 1)]^{1/2} N_{Q\pm 1}^{(K)}$$

or, that $\mathbf{N}^{(K)}$ is a spherical tensor of rank K in J space. This demonstrates that spherical tensors of ranks k and k' can be coupled to form a spherical tensor of rank K in exactly the same way as angular momenta of magnitudes k and

k' can be coupled to a total momentum of K; this also holds true for the coupling of tensors in two different spaces to form a tensor in the combined space. It is traditional[19] to define a special scalar product when two spherical tensors of integer rank k are coupled to a tensor of rank 0. Thus

$$\mathbf{T}^{(k)} \cdot \mathbf{U}^{(k)} = (-1)^k [k]^{1/2} (\mathbf{T}^{(k)} \mathbf{U}^{(k)})^{(0)}$$
$$= \sum_q (-1)^q T_q^{(k)} U_{-q}^{(k)}. \quad \text{(II-18)}$$

The Wigner-Eckart theorem (Section III-8) can be used to remove the dependence of matrix elements of spherical tensors on the magnetic quantum numbers

$$\langle JM_J | T_q^{(k)} | J'M_{J'} \rangle = (-1)^{J-M_J} \begin{pmatrix} J & k & J' \\ -M_J & q & M_{J'} \end{pmatrix} \langle J \| T^k \| J' \rangle \quad \text{(II-19)}$$

where $\langle J \| T^{(k)} \| J' \rangle$ is called a reduced matrix element, and is independent of M_J, q, and $M_{J'}$. Some common reduced matrix elements are given by

$$\langle j_1 \| j \| j_2 \rangle = \delta(j_1, j_2)[j_1(j_1+1)(2j_1+1)]^{1/2}$$
$$\langle l \| C^{(k)} \| l' \rangle = (-1)^l [l, l']^{1/2} \begin{pmatrix} l & k & l' \\ 0 & 0 & 0 \end{pmatrix}. \quad \text{(II-20)}$$

A particularly useful tensor operator, $w^{(\kappa k)}$, having rank κ in spin space and k in orbital space, has been introduced by Judd.[40] This operator is defined through its reduced matrix elements

$$\langle \tfrac{1}{2} l \| w^{(\kappa k)} \| \tfrac{1}{2} l' \rangle = \delta(l, l')[\kappa, k]^{1/2}. \quad \text{(II-21)}$$

Of particular interest to us are expressions for reduced matrix elements of $\mathbf{N}^{(K)}$ in terms of reduced matrix elements of $\mathbf{T}^{(k)}$ and $\mathbf{U}^{(k')}$. The simplest such relationship results from an uncoupling of $\mathbf{T}^{(k)}$ and $\mathbf{U}^{(k')}$;

$$\langle \gamma J \| N^{(K)} \| \gamma' J' \rangle = [K]^{1/2} (-1)^{J+J'+K} \sum_{J'',\gamma''} \begin{Bmatrix} k' & K & k \\ J & J'' & J' \end{Bmatrix}$$
$$\times \langle \gamma J \| T^{(k)} \| \gamma'' J'' \rangle \langle \gamma'' J'' \| U^{(k')} \| \gamma' J' \rangle \quad \text{(II-22)}$$

where γ includes all additional quantum numbers necessary to specify the states involved. Other important results are obtained when we assume the states J and J' are formed by coupling states j_1 and j_2, j_1' and j_2', respectively. Here, j_1 and j_1' are assumed to be angular momenta in one space, j_2 and j_2', angular momenta in another space. For instance, the first space could be a spin space, the second, an orbital space. Then, if $\mathbf{T}^{(k)}$ operates only in the first space and $\mathbf{U}^{(k')}$ in the second, we have

$$\langle \gamma j_1 j_2 J \| (\mathbf{T}^{(k)} \mathbf{U}^{(k')})^{(K)} \| \gamma' j'_1 j'_2 J' \rangle = \sum_{\gamma''} [J, K, J']^{\frac{1}{2}} \begin{Bmatrix} j_1 & j'_1 & k \\ j_2 & j'_2 & k' \\ J & J' & K \end{Bmatrix}$$
$$\times \langle \gamma j_1 \| T^{(k)} \| \gamma'' j''_1 \rangle \langle \gamma'' j''_2 \| U^{(k')} \| \gamma' j'_2 \rangle. \quad \text{(II-23)}$$

This equation has three special cases of particular interest. In the first, $K = 0, k = k'$, and

$$\langle \gamma j_1 j_2 JM | \mathbf{T}^{(k)} \cdot \mathbf{U}^{(k)} | \gamma' j'_1 j'_2 J'M' \rangle = (-1)^{j_1' + j_2 + J} \delta(J, J') \delta(M, M')$$
$$\times \begin{Bmatrix} j'_1 & j_2 & J \\ j_2 & j_1 & k \end{Bmatrix} \sum_{\gamma''} \langle \gamma j_1 \| T^{(k)} \| \gamma'' j''_1 \rangle \langle \gamma'' j''_2 \| U^{(k)} \| \gamma' j'_2 \rangle. \quad \text{(II-24)}$$

In the second case, we set $k' = 0$, $k = K$, and

$$\langle \gamma j_1 j_2 J \| T^{(k)} \| \gamma' j'_1 j'_2 J' \rangle = \delta(j_2, j'_2)(-1)^{j_1 + j_2 + J' + k}$$
$$\times [J, J']^{\frac{1}{2}} \begin{Bmatrix} J & k & J' \\ j'_1 & j_2 & j_1 \end{Bmatrix} \langle \gamma j_1 \| T^{(k)} \| \gamma' j'_1 \rangle. \quad \text{(II-25)}$$

For the final case, we set $k = 0$, $k' = K$ and

$$\langle \gamma j_1 j_2 J \| U^{(k')} \| \gamma' j'_1 j'_2 J' \rangle = \delta(j_1, j'_1)(-1)^{j_1' + j_2' + J + k'}$$
$$\times [J, J']^{\frac{1}{2}} \begin{Bmatrix} J & k' & J' \\ j'_2 & j_1 & j_2 \end{Bmatrix} \langle \gamma j_2 \| U^{(k')} \| \gamma' j'_2 \rangle. \quad \text{(II-26)}$$

5. SECOND-QUANTIZATION TECHNIQUES

It has been found that some of the formalisms of second quantization can be applied very profitably to atomic calculations. A thorough discussion of these applications has been given by Judd.[43] We briefly review here some of the results of Judd's work which will be of use to us in our study of hyperfine structure.

The creation operator $a_\alpha^\dagger$ is defined by the equation

$$a_\alpha^\dagger |0\rangle = |\alpha\rangle \quad \text{(II-27)}$$

where $|0\rangle$ is the vacuum state, and α represents the quantum numbers $(n l m_s m_l)$ of a single non-relativistic atomic electron. We wish to define the creation operators such that the usual determinental product state $\{\alpha \beta \ldots \delta\}$ is identical to a product of creation operators acting on the vacuum state:

$$\{\alpha \beta \ldots \delta\} = a_\alpha^\dagger a_\beta^\dagger \ldots a_\delta^\dagger |0\rangle.$$

If this equivalence is to hold, the creation operators must anticommute since

$$\{\alpha \beta \ldots \delta\} = -\{\beta \alpha \ldots \delta\} = -a_\beta^\dagger a_\alpha^\dagger \ldots a_\delta^\dagger |0\rangle.$$

Thus, we must have

$$a_\alpha^\dagger a_\beta^\dagger + a_\beta^\dagger a_\alpha^\dagger \equiv [a_\alpha^\dagger, a_\beta^\dagger]_+ = 0. \quad \text{(II-28)}$$

It is useful to note that when $\alpha = \beta$, Eq. (II-28) becomes

$$a_\alpha^\dagger a_\alpha^\dagger = -a_\alpha^\dagger a_\alpha^\dagger;$$

this implies that any operation in which two identical creation operators appear side-by-side must vanish identically.

The annihilation operator a_α is defined as the operator which destroys the state $|\alpha\rangle$:

$$a_\alpha |\alpha\rangle = |0\rangle. \quad \text{(II-29)}$$

This operator is the adjoint of the creation operator; thus

$$\{\alpha\beta \ldots \delta\}^* = \langle 0| a_\delta \ldots a_\beta a_\alpha.$$

Taking adjoints of Eq. (II-28), we find that the annihilation operators must also anticommute:

$$[a_\alpha, a_\beta]_+ = 0. \quad \text{(II-30)}$$

As a consequence of Eq. (II-30), it is clear that any operation containing two identical annihilation operators side-by-side must also vanish identically.

Finally, we require that two states formed by different sets of creation operators be orthonormal. For example, we require that

$$\int \{\alpha\beta\}^* \{\gamma\delta\} \, d\tau = \langle 0| a_\beta a_\alpha a_\gamma^\dagger a_\delta^\dagger |0\rangle$$

$$= \delta(\alpha, \gamma) \delta(\beta, \delta) - \delta(\alpha, \delta) \delta(\beta, \gamma).$$

This condition will be satisfied if

$$[a_\alpha^\dagger, a_\beta]_+ = \delta(\alpha, \beta) \quad \text{(II-31)}$$

and

$$a_\beta |0\rangle = 0.$$

Operators take on a special structure in the second quantization formalism. A common operator in atomic calculations is given by a sum of single particle operators

$$F = \sum_i f_i; \quad \text{(II-32a)}$$

this operator becomes in the second quantization formalism

$$F = \sum_{\alpha,\beta} a_\alpha^\dagger \langle \alpha | f | \beta \rangle a_\beta. \quad \text{(II-32b)}$$

Another important operator is described by a sum of two-body operators

$$G = \sum_{i<j} g_{ij}; \tag{II-33a}$$

this operator becomes in the second quantized representation

$$G = \frac{1}{2} \sum_{\alpha,\beta,\gamma,\delta} a_\alpha^\dagger a_\beta^\dagger \langle \alpha_1 \beta_2 | g_{12} | \gamma_1 \delta_2 \rangle a_\delta a_\gamma, \tag{II-33b}$$

where the subscripts in the matrix element refer to the number of the electron, making it clear that in the ket, for instance, electron 1 is in the state $|\gamma\rangle$.

6. GRAPHS

In the second quantization formalism, Feynman graphs[44] are often used to depict an interaction or series of interactions. These graphs can be quite useful in aiding one to visualize interactions and in helping to determine if all possible interactions have been considered. A few examples of Feynman graphs can perhaps best illustrate the procedure. The graphs

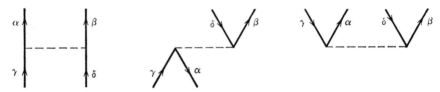

all illustrate examples of operators of the type

$$a_\alpha^\dagger a_\beta^\dagger \langle \alpha_1 \beta_2 | g_{12} | \gamma_1 \delta_2 \rangle a_\delta a_\gamma.$$

In the first graph on the left, α, β, γ, and δ are states in partially filled shells of an atom. In the second, α and δ are states in filled shells, β and γ are states in partially filled shells. (A filled nl shell is defined as one in which there are $4l + 2$ electrons. The Pauli principle forbids there being any more electrons in this shell. See Section III-3.) In the last diagram, γ and δ are states in filled shells, α and β are states in partially filled shells. One body operators

$$a_\alpha^\dagger \langle \alpha | f | \beta \rangle a_\beta$$

are expressed in terms of graphs of the type

In the first graph on the left, both states are in partially filled shells; in the second, β is in a filled shell, and α is in a partially filled shell.

The basic rules of graph-drawing should be clear from these examples. Arrows point up for lines associated with electrons in partially filled shells; these lines are called particle lines. Arrows point down for lines associated with electrons in filled shells; these lines are called hole lines. Interactions are designated by dotted lines. Interaction lines for two body interactions begin and end on solid lines; interaction lines for one body interactions begin at a solid line and end on a symbol representing the one body interaction. At each junction of an interaction line with solid lines, one solid line must have arrows pointing toward the junction and the other line, arrows pointing away from the junction. Alternative ground-rules exist for the drawing of graphs, but these appear to be the most useful for interactions in complex atoms.

An example can be given at this point to demonstrate the use of methods of second quantization and of graphical techniques. Let us consider the effect of operating on a state $|\psi_0\rangle$ with the Coulomb operator $\sum_{i<j} e^2/r_{ij}$. We assume that the state $|\psi_0\rangle$ contains a filled $n'l'$ shell and a partially filled nl shell. We wish to discuss only the effects of the terms in the second quantized form of the Coulomb operator which contain one creation and one annihilation operator referring to states in the $n'l'$ shell with the two remaining operators referring to states in the nl shell. If we designate the former operators by the symbol b and the latter by the symbol a, the terms of interest in the Coulomb operator can be written

$$\sum \left[a_\alpha^\dagger b_\beta^\dagger \langle \alpha_1 \beta_2 | \frac{e^2}{r_{12}} | \delta_1 \gamma_2 \rangle b_\gamma a_\delta + a_\alpha^\dagger b_\beta^\dagger \langle \alpha_1 \beta_2 | \frac{e^2}{r_{12}} | \gamma_1 \delta_2 \rangle a_\delta b_\gamma \right] \quad \text{(II-34)}$$

where the sums are now over only magnetic quantum numbers. Because the shell $n'l'$ is filled in $|\psi_0\rangle$, $b_\beta^\dagger|\psi_0\rangle = 0$. Thus our first task is to work the creation operators $b_\beta^\dagger$ in expression (II-34) to the right using the anticommutation relations of Eq. (II-31). When applied to $|\psi_0\rangle$, the only surviving terms in (II-34) can then be written

$$\sum \left[a_\alpha^\dagger \langle \alpha_1 \beta_2 | \frac{e^2}{r_{12}} | \delta_1 \gamma_2 \rangle a_\delta \, \delta(\beta, \gamma) - a_\alpha^\dagger \langle \alpha_1 \beta_2 | \frac{e^2}{r_{12}} | \gamma_1 \delta_2 \rangle a_\delta \, \delta(\beta, \gamma) \right]. \quad \text{(II-35)}$$

A delta function $\delta(\beta, \gamma)$ is represented graphically by joining the lines labeled β and γ. Thus the two terms above can be described by means of graphs as

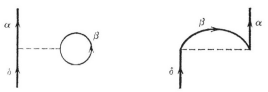

The diagram on the left is of a direct interaction; the diagram on the right, an exchange interaction. In this example, the $n'l'$ shell is said to be passive.

The above interaction can be described in terms of an effective one-body operator acting in the nl shell only. This can be seen in the expression (II-35), which can be interpreted as the second quantized form of a one-body operator f_{eff} having matrix elements

$$\langle \alpha | f_{\text{eff}} | \delta \rangle = \sum_{\beta} \left[\langle \alpha_1 \beta_2 | \frac{e^2}{r_{12}} | \delta_1 \beta_2 \rangle - \langle \alpha_1 \beta_2 | \frac{e^2}{r_{12}} | \beta_1 \delta_2 \rangle \right].$$

The one-body nature of the interaction is also clear from the graphs above, which have only two free lines, a characteristic of graphs describing one-body interactions.

III. Atomic Structure

1. INTRODUCTION

There are many excellent and detailed discussions of atomic structure.[32,40,45] Our approach to the question of atomic structure shall differ from most of these, however, in that we shall discuss the problem from the standpoint of relativistic electron theory. This approach is necessitated by our desire to consider hyperfine structure as a basically relativistic phenomena.

A major objection which can be raised to an approach of this type is that no exact Hamiltonian exists in closed form which describes the interaction of two relativistic electrons. This objection is certainly a valid one; however, we can find an approximate two-electron Hamiltonian by expanding in powers of v/c the expression for a one-photon exchange between two relativistic electrons which can be obtained by utilizing field theory.[46] In this manner, the leading terms in the interaction between two electrons are found to be

$$\frac{e^2}{r_{12}} - \left[\frac{e^2}{2}\frac{\boldsymbol{\alpha}_1 \cdot \boldsymbol{\alpha}_2}{r_{12}} + \frac{e^2}{2}\frac{(\boldsymbol{\alpha}_1 \cdot \mathbf{r}_{12})(\boldsymbol{\alpha}_2 \cdot \mathbf{r}_{12})}{r_{12}^3}\right] + O\left(\frac{v}{c}\right)^3 \quad \text{(III-1)}$$

where $\boldsymbol{\alpha}$ is a four-by-four matrix related to the two-by-two Pauli spin matrices

$$\sigma_x = \begin{pmatrix} 0 & 1 \\ 1 & 0 \end{pmatrix}, \quad \sigma_y = \begin{pmatrix} 0 & -i \\ i & 0 \end{pmatrix}, \quad \sigma_z = \begin{pmatrix} 1 & 0 \\ 0 & -1 \end{pmatrix},$$

by

$$\boldsymbol{\alpha} = \begin{pmatrix} 0 & \sigma \\ \sigma & 0 \end{pmatrix}$$

The term in brackets in (III-1) is known as the Breit interaction.[47] Although (III-1) is only an expansion in powers of v/c of the lowest order interaction between two electrons, we shall show that it contains all of the interactions which are, at least at the present time, necessary in order to interpret electronic structure of complex atoms.

In much the same spirit, we shall consider the one electron part of the many electron Hamiltonian to be given by the Dirac equation:[48]

$$H = \boldsymbol{\alpha} \cdot c\mathbf{p} + \beta mc^2 - \frac{Ze^2}{r}, \qquad \text{(III-2)}$$

where

$$\beta = \begin{pmatrix} 1 & 0 & 0 & 0 \\ 0 & 1 & 0 & 0 \\ 0 & 0 & -1 & 0 \\ 0 & 0 & 0 & -1 \end{pmatrix}.$$

This means, of course, that no interactions of the electrons with the vacuum will be included,[46] and no "Lamb shift" type of corrections can come out of our discussion. However, the error introduced by neglecting shifts of this type is generally much smaller than the errors resulting from the many approximations which must be used in dealing with complex atoms. With these approximations, the Hamiltonian for N electrons moving in the field of a point nucleus of charge Z is

$$H = \sum_i \left(\boldsymbol{\alpha}_i \cdot c\mathbf{p}_i + \beta m_i c^2 - \frac{Ze^2}{r_i} \right) + \sum_{i<j} \left(\frac{e^2}{r_{ij}} - \tfrac{1}{2} e^2 \frac{\boldsymbol{\alpha}_i \cdot \boldsymbol{\alpha}_j}{r_{ij}} - \tfrac{1}{2} e^2 \frac{(\boldsymbol{\alpha}_i \cdot \mathbf{r}_{ij})(\boldsymbol{\alpha}_j \cdot \mathbf{r}_{ij})}{r_{ij}^3} \right) \qquad \text{(III-3)}$$

where the charge of an electron is $-e$.

Equation (III-3) describes a very complex many-body problem which cannot be solved exactly in the general case; perturbation theory must therefore be utilized to obtain approximate solutions. In order to use perturbation theory, we make the "central field" approximation; that is, each electron is assumed to move in a spherically symmetric field which is produced by the nuclear field and the spherically averaged fields of all other electrons. The exact form of the central field potential will be discussed in Chapter V. The Hamiltonian (III-3) can then be written as a sum of a large part H_0, which is itself a sum of single particle Hamiltonians, and a smaller part H_1, which contains both one and two body operators. Thus

$$H = H_0 + H_1 \qquad \text{(III-4a)}$$

where

$$H_0 = \sum_i \left(\boldsymbol{\alpha}_i \cdot c\mathbf{p}_i + \beta m_i c^2 - \frac{Ze^2}{r_i} + U(r_i) \right) = \sum_i H_c(i) \qquad \text{(III-4b)}$$

$$H_1 = -\sum_i U(r_i) + \sum_{i<j} \left(\frac{e^2}{r_{ij}} - \tfrac{1}{2} e^2 \frac{\boldsymbol{\alpha}_i \cdot \boldsymbol{\alpha}_j}{r_{ij}} - \tfrac{1}{2} e^2 \frac{(\boldsymbol{\alpha}_i \cdot \mathbf{r}_{ij})(\boldsymbol{\alpha}_i \cdot \mathbf{r}_{ij})}{r_{ij}^3} \right). \qquad \text{(III-4c)}$$

We shall address ourselves to the problems of obtaining eigenfunctions of H_0 and understanding their properties in the next four sections. In Section III-6, we shall turn to the evaluation of the perturbation term H_1.

2. THE DIRAC EQUATION

The H_0 is simply a sum of N single particle Dirac equations for an electron in a central field; the eigenfunction for H_0, Ψ_0, can therefore be expressed as a product of N single particle wavefunctions ψ_i. In this section, we confine our attention to finding ψ_i, the eigenfunction of the ith Dirac Hamiltonian, $H_c(i)$. Because only the Hamiltonian for the ith electron is discussed, the index i can be suppressed for the remainder of this section without causing confusion.

We shall show that the eigenfunction ψ can be written as a product of a function which depends on the radius, and a function which depends on the angular coordinates θ and ϕ. First, it is convenient to define

$$\psi = \begin{pmatrix} \mu \\ \varepsilon \end{pmatrix} \tag{III-5}$$

where μ and ε are one-column matrices of two rows. A hint as to the explicit form of ψ can be obtained by noting that H_c does not commute either with the orbital angular momentum $\mathbf{L}$ ($L_1 = -i(x_2(\partial/\partial x_3) - x_3(\partial/\partial x_2))$), for example

$$[L_x, H_c] = [L_x, \boldsymbol{\alpha} \cdot c\mathbf{p}] = ic(\alpha_y p_z - \alpha_z p_y);$$

or with $\mathbf{S} = \tfrac{1}{2}\begin{pmatrix} \sigma & 0 \\ 0 & \sigma \end{pmatrix}$, for example

$$[S_x, H_c] = [S_x, \boldsymbol{\alpha} \cdot c\mathbf{p}] = -ic(\alpha_y p_z - \alpha_z p_y).$$

However, H_c does commute with the vector sum of these two operators, $\mathbf{J} = \mathbf{L} + \mathbf{S}$, and consequently also with the operator $\mathbf{J}^2$. Thus, if we choose the z axis as axis of quantization, ψ can be constructed so as to be an eigenfunction not only of H_c, but also of $\mathbf{J}^2$ and J_z.

The eigenfunctions ψ must in the non-relativistic limit reduce to the familiar eigenfunctions obtained for a non-relativistic electron moving in a central field in the presence of spin-orbit coupling. We are led, therefore, to try the trial forms for μ and ε:

$$\mu = \frac{F(r)}{r} \mathscr{Y}_{ljm}$$

$$\varepsilon = \frac{iG(r)}{r} \mathscr{Y}_{l'jm} \tag{III-6}$$

where $F(r)$ and $G(r)$ are functions of radius only, and

$$\mathcal{Y}_{ljm} = \sum_{\substack{m_s \\ m_l}} (-1)^{l-\frac{1}{2}-m}[j]^{\frac{1}{2}} \begin{pmatrix} \frac{1}{2} & l & j \\ m_s & m_l & -m \end{pmatrix} Y_{lm_l}(\theta, \phi) \chi_{m_s}^{\frac{1}{2}}$$

is the angular portion of the non-relativistic wavefunction describing an electron for which the spin is coupled to the orbital momentum, $\mathbf{l}$, to form a total angular momentum, $\mathbf{j}$. $Y_{lm_l}(\theta, \phi)$ is a spherical harmonic ($0 \leq \theta \leq \pi$, $0 \leq \phi \leq 2\pi$) satisfying the eigenvalue equations

$$L_z Y_{lm} = m Y_{lm},$$
$$\mathbf{L}^2 Y_{lm} = l(l+1) Y_{lm}$$

and $\chi_{m_s}^{\frac{1}{2}}$ is a two component spinor satisfying the eigenvalue equations

$$\tfrac{1}{2}\sigma_z \chi_{m_s}^{\frac{1}{2}} = m_s \chi_{m_s}^{\frac{1}{2}},$$
$$\tfrac{1}{4}\sigma^2 \chi_{m_s}^{\frac{1}{2}} = \tfrac{1}{2}(\tfrac{1}{2}+1)\chi_{m_s}^{\frac{1}{2}}.$$

With the form given above for μ and ε, one can easily show that ψ satisfies the eigenvalue equations

$$J_z \psi = m\psi$$
$$\mathbf{J}^2 \psi = j(j+1)\psi. \tag{III-7}$$

The parity operator, P, also commutes with H_c. In the four-component theory, P can be written as βP_0, where P_0 reverses the sign of the space coordinates only. Thus

$$P\begin{pmatrix}\mu \\ \varepsilon\end{pmatrix} = P_0 \begin{pmatrix}\mu \\ -\varepsilon\end{pmatrix}.$$

Obviously, if ψ is to have a definite parity, μ and ε must differ in parity. Since the parity of Y_{lm} is $(-1)^l$, both the parity condition and the condition imposed by the 3-j symbol of Eq. (III-6) will be satisfied if $l' = l \pm 1$ as $j = l \pm \tfrac{1}{2}$.

We now consider the eigenvalue equation

$$H_c \psi = E\psi; \tag{III-8}$$

using the form of ψ given by Eq. (III-5), this can be written as

$$\begin{pmatrix} c\boldsymbol{\sigma} \cdot \mathbf{p}\varepsilon + \left(mc^2 + U(r) - \dfrac{Ze^2}{r}\right)\mu \\ c\boldsymbol{\sigma} \cdot \mathbf{p}\mu - \left(mc^2 - U(r) + \dfrac{Ze^2}{r}\right)\varepsilon \end{pmatrix} = \begin{pmatrix} E\mu \\ E\varepsilon \end{pmatrix}.$$

Equation (III-8) can therefore be expressed in terms of the coupled equations:

$$\left(mc^2 + U(r) - \frac{Ze^2}{r} - E\right)\frac{F(r)}{r}\mathcal{Y}_{ljm} = -ic\boldsymbol{\sigma}\cdot\mathbf{p}\,\frac{G(r)}{r}\mathcal{Y}_{l'jm}$$
$$\left(mc^2 - U(r) + \frac{Ze^2}{r} + E\right)\frac{G(r)}{r}\mathcal{Y}_{l'jm} = -ic\boldsymbol{\sigma}\cdot\mathbf{p}\,\frac{F(r)}{r}\mathcal{Y}_{ljm}. \tag{III-9}$$

These coupled equations can be simplified by manipulating the right-hand sides. First, we consider

$$\boldsymbol{\sigma}\cdot\mathbf{C}^{(1)}\mathcal{Y}_{ljm} = \sum_{\bar{l}}\langle\mathcal{Y}_{\bar{l}jm}|\boldsymbol{\sigma}\cdot\mathbf{C}^{(1)}|\mathcal{Y}_{ljm}\rangle\mathcal{Y}_{\bar{l}jm}$$

$$= \sum_{\bar{l}}(-1)^{\frac{1}{2}+l+j}\begin{Bmatrix}\frac{1}{2} & l & j\\ \bar{l} & \frac{1}{2} & 1\end{Bmatrix}\langle\frac{1}{2}\|\sigma\|\frac{1}{2}\rangle\langle\bar{l}\|C^{(1)}\|l\rangle\mathcal{Y}_{\bar{l}jm}$$

$$= \mathcal{Y}_{l'jm}.$$

In addition, we need to carry out a recoupling

$$((\sigma p)^{(0)}(\sigma C^{(1)})^{(0)})^{(0)} = \sum_k \begin{Bmatrix}1 & 1 & 0\\ 1 & 1 & 0\\ k & k & 0\end{Bmatrix}((\sigma\sigma)^{(k)}(pC^{(1)})^{(k)})^{(0)}[k]$$

$$= \sum_k \frac{[k]^{\frac{1}{2}}}{[1]}((\sigma\sigma)^{(k)}(pC^{(1)})^{(k)})^{(0)}.$$

This can be written, using the identities $(\sigma\sigma)^{(0)} = -[1]^{-\frac{1}{2}}(\boldsymbol{\sigma}\cdot\boldsymbol{\sigma}) = -\sqrt{3}$; $(\sigma\sigma)^{(1)} = i(\boldsymbol{\sigma}\times\boldsymbol{\sigma})/\sqrt{2} = -\sqrt{2}\boldsymbol{\sigma}$; and $(\sigma\sigma)^{(2)} = 0$; as

$$(\boldsymbol{\sigma}\cdot\mathbf{p})(\boldsymbol{\sigma}\cdot\mathbf{C}^{(1)}) = \mathbf{p}\cdot\mathbf{C}^{(1)} - i\boldsymbol{\sigma}\cdot\mathbf{C}^{(1)}\times\mathbf{p} = \mathbf{p}\cdot\mathbf{C}^{(1)} - i\hbar\,\frac{\boldsymbol{\sigma}\cdot\mathbf{L}}{r}.$$

We have also used Eq. (II-18) and $\mathbf{a}^{(1)}\times\mathbf{b}^{(1)} = -i\sqrt{2}(\mathbf{a}^{(1)}\mathbf{b}^{(1)})^{(1)}$ in order to obtain the above result. Combining these two results, one finds that

$$\boldsymbol{\sigma}\cdot\mathbf{p}\,\mathcal{Y}_{l'jm} = -\frac{i\hbar}{r}(2+\boldsymbol{\sigma}\cdot\mathbf{L})\mathcal{Y}_{ljm}.$$

Then

$$\boldsymbol{\sigma}\cdot\mathbf{p}\left(\frac{G}{r}\mathcal{Y}_{l'jm}\right) = -i\hbar\,\frac{d}{dr}\left(\frac{G}{r}\right)\boldsymbol{\sigma}\cdot\mathbf{C}^{(1)}\mathcal{Y}_{l'jm} + \frac{G}{r}\boldsymbol{\sigma}\cdot\mathbf{p}\,\mathcal{Y}_{l'jm}$$

$$= -i\hbar\left[\frac{1}{r}\frac{dG}{dr} + \frac{G}{r^2}(1+\boldsymbol{\sigma}\cdot\mathbf{L})\right]\mathcal{Y}_{ljm}.$$

26 Atomic Structure

The quantity κ is defined by the equation

$$\boldsymbol{\sigma} \cdot \mathbf{L} \mathcal{Y}_{ljm} = (\kappa - 1)\mathcal{Y}_{ljm}$$
$$= [j(j+1) - l(l+1) - \tfrac{3}{4}]\mathcal{Y}_{ljm} \qquad \text{(III-10)}$$

or

$$\kappa = (-1)^{j+l-\frac{1}{2}} \frac{[j]}{2}.$$

Noting, finally, that $(1 + \boldsymbol{\sigma} \cdot \mathbf{L})\mathcal{Y}_{ljm} = -(1 + \boldsymbol{\sigma} \cdot \mathbf{L})\mathcal{Y}_{l'jm}$, we can put Eqs. (III-9) into the form

$$\left(\frac{d}{dr} + \frac{\kappa}{r}\right)G(r) = \frac{-1}{\hbar c}\left(-E + mc^2 + U(r) - \frac{Ze^2}{r}\right)F(r)$$
$$\left(\frac{d}{dr} - \frac{\kappa}{r}\right)F(r) = \frac{-1}{\hbar c}\left(E + mc^2 - U(r) + \frac{Ze^2}{r}\right)G(r). \qquad \text{(III-11)}$$

In order that ψ be normalized, we must also require that F and G satisfy the integral equation

$$\int (F(r)^2 + G(r)^2)\, dr = 1. \qquad \text{(III-12)}$$

For a given κ, Eqs. (III-11) and (III-12) do not uniquely define the functions of F and G and the value of the eigenvalue E. There are, in fact, an infinity of possible solutions. These solutions can be labeled by an integer n, called the principal quantum number; n is constrained to be greater than or equal to $l+1$. To fully specify a set of solutions, therefore, we must affix the quantum numbers n, l and j (or κ), e.g. F_{nlj}, G_{nlj}, and E_{nlj}.

It is interesting to investigate the equations which F and G satisfy in the non-relativistic limit. In this case, we may write

$$E \simeq E_0 + mc^2$$

where E_0 is the sum of the usual non-relativistic kinetic and potential energies of the electron. The second of Eqs. (III-11) can then be written

$$G \simeq -\hbar c \left(E_0 + 2mc^2 - U(r) + \frac{Ze^2}{r}\right)^{-1}\left(\frac{d}{dr} - \frac{\kappa}{r}\right)F$$
$$\simeq -\frac{\mu_0}{e}\left(\frac{d}{dr} - \frac{\kappa}{r}\right)F \qquad \text{(III-13)}$$

where $\mu_0 = e\hbar/2mc$. The second expression on the right above results from an expansion of the first in powers of $(E_0 - U(r) + Ze^2/r)/2mc^2 \simeq (v/c)^2$, retaining only the lowest order term. (See, however, the Appendix.) This expression can be inserted into the first of Eqs. (III-11) in order to obtain an equation in F only:

$$\left[\frac{-\hbar^2}{2m}\left(\frac{d^2}{dr^2} - \frac{l(l+1)}{r^2}\right) + U(r) - \frac{Ze^2}{r}\right]F = E_0 F.$$

This is simply the non-relativistic radial Schrodinger equation for an electron in a central field; note that F is independent of j in this limit and can be specified by the quantum numbers n and l only. Thus to an accuracy of $(v/c)^2$, F is the ordinary radial wavefunction, and G is roughly $\hbar/2mc$ as large. In keeping with this result, μ is known as the large component of ψ; ε, the small component.

3. CREATION AND ANNIHILATION OPERATORS FOR RELATIVISTIC ELECTRONS

The creation operator $q^\dagger(nljm)$ is defined by the equation

$$q^\dagger(nljm)|0\rangle = \begin{pmatrix} \dfrac{F_{nlj}}{r}\mathcal{Y}_{ljm} \\ \dfrac{iG_{nlj}}{r}\mathcal{Y}_{l'jm} \end{pmatrix} \quad \text{(III-14)}$$

where $|0\rangle$ is the vacuum state. The state on the right-hand side of Eq. (III-14) will often be written as $|nljm\rangle$. Taking adjoints of both sides of Eq. (III-14) leads to the equation

$$\langle 0|q(nljm) = \langle nljm|.$$

Finally, the equation

$$\langle nljm|nljm\rangle = 1 = \langle 0|q(nljm)|nljm\rangle$$

demonstrates that $q(nljm)$ must be the operator which annihilates the state $|nljm\rangle$, that is

$$q(nljm)|nljm\rangle = |0\rangle. \quad \text{(III-15)}$$

A many-body state such as Ψ_0 must, because of the Pauli exclusion principle,[49] be expressed as an antisymmetrized product of single particle wavefunctions. Traditionally, such a wavefunction is written as a determinantal product state $\{\alpha\beta\ldots\delta\}$ where α, and so on, stand for single particle states with quantum numbers $(n_\alpha l_\alpha j_\alpha m_\alpha)$, and so on. As in Section II-5, such an antisymmetrized state can also be written as a product of creation operators acting on the vacuum

$$q^\dagger(\alpha)q^\dagger(\beta)\ldots q^\dagger(\delta)|0\rangle$$

if the creation operators are made to satisfy the anticommutation rules

$$[q^\dagger(\alpha), q^\dagger(\beta)]_+ = 0. \qquad \text{(III-16a)}$$

Taking adjoints of this equation, we find that the annihilation operators also anticommute:

$$[q(\alpha), q(\beta)]_+ = 0. \qquad \text{(III-16b)}$$

Finally, if the antisymmetric states defined by creation operators are to be normalized, we must require

$$[q^\dagger(\alpha), q(\beta)]_+ = \delta(\alpha, \beta). \qquad \text{(III-16c)}$$

By writing the operator **J** in second quantized form:

$$\mathbf{J} = \sum_{\alpha, \beta} q^\dagger(\alpha)(\alpha|\mathbf{j}|\beta)q(\beta)$$

we can investigate the commutation relations of the creation and annihilation operators with the total angular momentum, **J**. One can easily show that

$$[J_z, q^\dagger(nljm)] = m q^\dagger(nljm)$$

$$[J_\pm, q^\dagger(nljm)] = [j(j+1) - m(m \pm 1)]^{1/2} q^\dagger(nljm \pm 1).$$

These equations are equivalent to those used in Section II-4 to define spherical tensors of rank j: we find, therefore, that the operators $q^\dagger(nljm)$ for a fixed n, l, and j and with $-j \le m \le j$ form the components of a tensor of rank j, $\mathbf{q}^\dagger(nlj)$. An equivalent calculation performed on the annihilation operators shows that the operators $\tilde{q}(nljm) = (-1)^{j-m} q(nlj-m)$ also form the components of a tensor of rank j, $\mathbf{q}(nlj)$.

Although it is clear from the discussion of the previous section that l is not a good quantum number for relativistic electrons, it is convenient to define a relativistic sl state by the equation

$$|n\tfrac{1}{2}lm_s m_l\rangle = \sum_{j, m} [j]^{1/2}(-1)^{l-1/2-m} \begin{pmatrix} \tfrac{1}{2} & l & j \\ m_s & m_l & -m \end{pmatrix} |nljm\rangle. \qquad \text{(III-17)}$$

This equation can also be taken as the definition of $q^\dagger(n\tfrac{1}{2}lm_s m_l)$, the operator which creates the state on the left above, in terms of the operators $q^\dagger(nljm)$. Such an expansion is equivalent to ignoring the tensorial properties of the small components of the states $|nljm\rangle$. Because of this, Eq. (III-17) must be considered, in the relativistic extreme, to provide a definition rather than an equality; on the other hand, in the non-relativistic limit, when the small component of the wavefunction goes to zero, this equation is the usual one[40] for a transformation from $(slm_s m_l)$ states to $(sljm)$ states. Thus, we can justify the definition (III-17) by noting that it is reasonable and meaningful in the nonrelativistic limit.

Expressing Eq. (III-17) in terms of creation operators and taking adjoints, we further find

$$q(n\tfrac{1}{2}lm_s m_l) = \sum_{j,m}[j]^{1/2}(-1)^{l-1/2-m}\begin{pmatrix}\tfrac{1}{2} & l & j \\ m_s & m_l & -m\end{pmatrix}q(nljm). \quad \text{(III-18)}$$

Utilizing these definitions and carrying out sums over the 3-j symbols involved, we obtain the anticommutators

$$[q^\dagger(n\tfrac{1}{2}lm_s m_l),\ q^\dagger(n\tfrac{1}{2}l'm'_s m'_l)]_+ = 0 \quad \text{(III-19a)}$$

$$[q(n\tfrac{1}{2}lm_s m_l),\ q(n\tfrac{1}{2}l'm'_s m'_l)]_+ = 0 \quad \text{(III-19b)}$$

$$[q^\dagger(n\tfrac{1}{2}lm_s m_l),\ q(n\tfrac{1}{2}l'm'_s m'_l)]_+ = \delta(l,l')\delta(m_l,m'_l)\delta(m_s,m'_s). \quad \text{(III-19c)}$$

Further, it is straightforward to show that $q(n\tfrac{1}{2}lm_s m_l)$ defined above annihilates the state $|n\tfrac{1}{2}lm_s m_l\rangle$. We have, therefore, a set of fermion creation and annihilation operators for relativistic sl states.

By investigating commutators of the vectors **L** and **S**, defined by

$$L_\pm = \sum q^\dagger(nlm_s m_l \pm 1)[l(l+1) - m_l(m_l \pm 1)]^{1/2} q(nlm_s m_l),$$

$$S_\pm = \sum q^\dagger(nlm_s \pm \tfrac{1}{2} m_l)[\tfrac{3}{4} - m_s(m_s \pm 1)]^{1/2} q(nlm_s m_l),$$

$$L_z = \sum q^\dagger(nlm_s m_l)m_l q(nlm_s m_l),\quad S_z = \sum q^\dagger(nlm_s m_l)m_s q(nlm_s m_l),$$

with the (sl) creation and annihilation operators given above, one finds that the operators $q^\dagger(n\tfrac{1}{2}lm_s m_l)$ with n and l fixed, $-l \le m_l \le l$ and $-\tfrac{1}{2} \le m_s \le \tfrac{1}{2}$, form the components of a tensor $\mathbf{q}^\dagger(n\tfrac{1}{2}l)$ of rank l in orbital (L) space, $\tfrac{1}{2}$ in spin (S) space. Likewise, the operators $\tilde{q}(n\tfrac{1}{2}lm_s m_l) = (-1)^{l+1/2-m_s-m_l} q(n\tfrac{1}{2}l - m_s - m_l)$ form the components of another tensor, $\mathbf{q}(n\tfrac{1}{2}l)$, having these same properties in spin and orbital spaces.

We note that because of the Pauli principle (mathematically represented above by Eqs. (III-16a) and (III-19a)), Ψ_0 can contain no more than $2j+1$ electrons of the type $(nljm)$ for a fixed n, l and j. Likewise, for a given n and l, there can be no more than $(2s+1)(2l+1)$ electrons of the type $(n\tfrac{1}{2}lm_s m_l)$. The collection of configurations $(nlj)^{N_j}(0 \le N_j \le 2j+1)$ forms a *shell* in the (nlj) scheme; the collection $(n\tfrac{1}{2}l)^{N_l}(0 \le N_l \le (2s+1)(2l+1))$ forms a shell in the $(n\tfrac{1}{2}l)$ scheme. When $N_j = 2j+1$ or $N_l = 4l+2$, the shell is said to be closed.

Finally, before ending our discussion of creation and annihilation operators, let us consider an operator formed by coupling together two of these operators using their tensorial properties. We write:

$[\mathbf{q}^\dagger(n\tfrac{1}{2}l)\mathbf{q}(n'\tfrac{1}{2}l')]^{(\kappa k)}_{m_\kappa m_k}$

$$= \sum_{m_l, m_s} \begin{pmatrix} l & l' & k \\ m_{l_1} & -m_{l_2} & -m_k \end{pmatrix} \begin{pmatrix} \tfrac{1}{2} & \tfrac{1}{2} & \kappa \\ m_{s_1} & -m_{s_2} & -m_\kappa \end{pmatrix}$$
$$\times [\kappa, k]^{1/2}(-1)^{m_\kappa + m_k + l + \tfrac{1}{2} + m_{s_2} + m_{l_2}} q^\dagger(n\tfrac{1}{2}lm_{s_1}m_{l_1})q(n'\tfrac{1}{2}l'm_{s_2}m_{l_2}). \quad \text{(III-20)}$$

The right-hand side of this equation can be expressed as

$$-\sum_{m_s, m_l} q^\dagger(n\tfrac{1}{2}lm_{s_1}m_{l_1})(n\tfrac{1}{2}lm_{s_1}m_{l_1}|r^{(\kappa k)}_{m_\kappa m_k}(nl, n'l')|n'\tfrac{1}{2}l'm_{s_2}m_{l_2})q(n'\tfrac{1}{2}l'm_{s_2}m_{l_2})$$

where $\mathbf{r}^{(\kappa k)}$ is defined as a tensor of rank κ in spin space and k in orbital space, with reduced matrix element

$$(n_1\tfrac{1}{2}l_1\|r^{(\kappa k)}(n_\alpha l_\alpha, n_\beta l_\beta)\|n_2\tfrac{1}{2}l_2)$$
$$= \delta(n_1, n_\alpha)\,\delta(n_2, n_\beta)\,\delta(l_1, l_\alpha)\,\delta(l_2, l_\beta)[\kappa, k]^{1/2}. \quad \text{(III-21)}$$

By comparing the above result with the usual second quantized form for a one body operator (Section II-5), we see that $(\mathbf{q}^\dagger(n\tfrac{1}{2}l)\mathbf{q}(n'\tfrac{1}{2}l'))^{(\kappa k)}$ can be considered to be the second quantized form of the tensor operator

$$-\mathbf{R}^{(\kappa k)}(nl, n'l') = -\sum_i \mathbf{r}_i^{(\kappa k)}(nl, n'l').$$

As we shall show in the next few sections, the operators $\mathbf{R}^{(\kappa k)}$ play a very important role in the defining of states and the interpretation of the fine structure and hyperfine structure interactions.

4. GROUPS AND GROUP GENERATORS

The creation and annihilation operators introduced in the previous section for relativistic electrons of the types $(nljm)$ and $(n\tfrac{1}{2}lm_s m_l)$ satisfy anti-commutation relations which are identical to those satisfied by creation and annihilation operators for non-relativistic states having the same quantum numbers (Section II-5 and Lawson and Macfarlane).[50] All statements made in this section shall be based on these anticommutation relations, and so can be applied equally to both relativistic and non-relativistic electrons.

Before we consider these operators any further, we shall present a very brief review of the pertinent parts of group theory. For more complete discussion of group theory, we suggest the reader refer to for example Wigner,[51] Judd,[40] Hamermesh,[52] or Racah.[53] The results given below follow primarily the work of Judd.[40]

A *group* can be formed by any set of elements satisfying the following conditions:

1. Any two elements in the set can be combined to give a third element in the set, that is,

$$AB = C$$

where A, B, and C are elements of the set.

2. The triple product of elements in the set is well defined, that is,

$$A(BC) = (AB)C.$$

3. The set contains an identity element E which, for every element A in the set, has the property

$$AE = EA = A.$$

4. For every element A of the set there also exists in the set the inverse element A^{-1} having the property

$$AA^{-1} = A^{-1}A = E.$$

For concreteness, let us assume that the elements of our group are a set of transformations which carry the point $\mathbf{r}$ in an n dimensional space into the point $\mathbf{r}'$. Further, we assume that these transformations can be uniquely and completely defined by some set of p parameters, for example, the transformations can be written as

$$F(\mathbf{r}; a_1, a_2, \ldots a_p).$$

If we can start at any point $\mathbf{r}$ in the space and reach any other point in the space, $\mathbf{r}'$, by a continuous variation of the parameters a_i, the transformations F are said to form a *continuous* group. Lie showed[54] that the properties of such a group can be understood by studying transformations which are only infinitesimally different from the identity transformation. That is, all operators in the group can be expanded in the form

$$S_\alpha = 1 + \sum_\sigma \delta a^\sigma X_\sigma.$$

The X_σ are called the *infinitesimal operators*, or *generators*, of the group: they satisfy the relationships

$$[X_\alpha, X_\beta] = \sum_\gamma c(\alpha, \beta, \gamma) X_\gamma. \qquad \text{(III-22)}$$

Conversely, Lie has also shown that if a set of operators satisfy Eq. (III-22) then a group can be formed from the elements $1 + \sum \delta a^\sigma X_\sigma$. The elements X_σ and their commutators form a *Lie algebra*.

A *subgroup* is defined as some set of operators contained in a group which themselves satisfy the group requirements. Let us assume a subgroup having elements S_a; if, for all elements S_α contained in the full group, $S_\alpha S_a S_\alpha^{-1}$ is also an element of the subgroup, the subgroup is called *invariant*. Finally, a *simple* group is defined as one in which there exists no invariant subgroup other than the identity. It is with simple groups that we shall be concerned in the remainder of this section.

The infinitesimal operators (or sums of infinitesimal operators) can always be divided into two types,[55] H_i ($i = 1, 2, \ldots q$) and E_α, satisfying the equation

$$[H_i, H_j] = 0$$
$$[H_i, E_\alpha] = \alpha_i E_\alpha$$
$$[E_\alpha, E_\beta] = c(\alpha, \beta, \alpha + \beta) E_{\alpha+\beta} \quad (\beta \neq -\alpha) \quad \text{(III-23)}$$
$$[E_\alpha, E_{-\alpha}] = \sum \alpha_i H_i.$$

The numbers α_i can be considered to form components of a vector **α** in a q dimensional space called the *weight space*. The vector **α** is called a *root*.

Cartan[56] has shown that, with only five exceptions, the simple groups fall into four general classes. These classes are: special unitary groups, rotation groups in an even number of dimensions, rotation groups in an odd number of dimensions, and symplectic groups. By comparing a root figure of a simple group with the catalogue of root figures of simple groups given by Cartan,[56] van der Waerden[57] and Racah,[53] the group can easily be identified.

Let us now consider operators composed of the creation and annihilation operators for relativistic (*sl*) states with n and l held fixed. The set of operators

$$[q^\dagger(\alpha), q(\alpha)],$$
$$q^\dagger(\alpha)q(\beta), \quad (\alpha \neq \beta)$$
$$q^\dagger(\alpha)q^\dagger(\beta), \quad \text{(III-24a)}$$
$$q(\alpha)q(\beta),$$
$$q^\dagger(\alpha), q(\alpha),$$

where $(\alpha) = (n\tfrac{1}{2}lm_s m_l)$ and $(\beta) = (n\tfrac{1}{2}lm_{s'} m_{l'})$ are both allowed to run over all $4l + 2$ possible values of m_s and m_l, can easily be seen to form a set of operators having the properties of Eq. (III-22). Identifying the H operators as the set $H_\alpha = \tfrac{1}{2}[q^\dagger(\alpha), q(\alpha)]$, one finds

$$[H_\alpha, q^\dagger(\beta)] = \delta(\alpha, \beta) q^\dagger(\beta)$$
$$[H_\alpha, q(\beta)] = -\delta(\alpha, \beta) q(\beta)$$
$$[H_\alpha, q^\dagger(\beta) q^\dagger(\gamma)] = \{\delta(\alpha, \beta) + \delta(\alpha, \gamma)\} q^\dagger(\beta) q^\dagger(\gamma) \quad \text{(III-24b)}$$
$$[H_\alpha, q(\beta) q(\gamma)] = -\{\delta(\alpha, \beta) + \delta(\alpha, \gamma)\} q(\beta) q(\gamma)$$
$$[H_\alpha, q^\dagger(\beta) q(\gamma)] = \{\delta(\alpha, \beta) - \delta(\alpha, \gamma)\} q^\dagger(\beta) q(\gamma).$$

The root figure resulting from these equations quickly leads to the identification of the group formed by the operators (III-24) as the proper rotation group in $8l + 5$ dimensions, $R(8l + 5)$. A subgroup of this group can easily be found by discarding the single creation and annihilation operators, $q^\dagger(\alpha)$ and $q(\alpha)$, from the set (III-24a). The resulting group is another proper rotation group, $R(8l + 4)$.

Generators of subgroups of $R(8l + 4)$ can be described very concisely in terms of the operator $\mathbf{R}^{(\kappa k)}(nl, nl)$, which we shall abbreviate as $\mathbf{R}^{(\kappa k)}$. The $\mathbf{R}^{(\kappa k)}$'s commute among themselves according to the equation

$$[R^{(\kappa k)}_{\pi q}, R^{(\kappa' k')}_{\pi' q'}] = \sum (-1)^{1+\pi''+q''}\{(-1)^{\kappa+\kappa'+\kappa''+k+k'+k''} - 1\}$$
$$\times \begin{Bmatrix} \kappa & \kappa' & \kappa'' \\ s & s & s \end{Bmatrix} \begin{Bmatrix} k & k' & k'' \\ l & l & l \end{Bmatrix} \begin{pmatrix} \kappa & \kappa' & \kappa'' \\ \pi & \pi' & -\pi'' \end{pmatrix} \begin{pmatrix} k & k' & k'' \\ q & q' & -q'' \end{pmatrix}$$
$$\times [\kappa, \kappa', \kappa'', k, k', k'']^{1/2} R^{(\kappa'' k'')}_{\pi'' q''} \qquad \text{(III-25)}$$

where the sum is over all double primed quantities. Obviously, a group can be formed by all of the possible operators $\mathbf{R}^{(\kappa k)}$; because $\mathbf{R}^{(\kappa k)}$ can be expressed as a linear combination of pairs of operators $q^\dagger(\alpha)q(\beta)$, this group must be a subgroup of $R(8l + 4)$ amongst whose generators are all pairs of this type. The group formed by the operators $\mathbf{R}^{(\kappa k)}$ turns out to be the unitary group in $4l + 2$ dimensions, $U(4l + 2)$. A more interesting subgroup of $R(8l + 4)$ from the standpoint of atomic physics, however, is obtained by restricting the sum $\kappa + k$ to be odd. As can be seen from Eq. (III-25), if $\kappa + k$ and $\kappa' + k'$ are both odd, $\kappa'' + k''$ must also be odd. The group formed by this set of generators is the symplectic group in $4l + 2$ dimensions, $Sp(4l + 2)$.

Another very useful subgroup of $R(8l + 4)$ is the $R(3)$ group having the generators

$$Q_+ = \left(\frac{[l]}{2}\right)^{1/2}(\mathbf{q}^\dagger(n\tfrac{1}{2}l)\mathbf{q}^\dagger(n\tfrac{1}{2}l))^{(00)}$$

$$Q_- = -\left(\frac{[l]}{2}\right)^{1/2}(\mathbf{q}(n\tfrac{1}{2}l)\mathbf{q}(n\tfrac{1}{2}l))^{(00)} \qquad \text{(III-26)}$$

$$Q_z = -\left(\frac{[l]}{8}\right)^{1/2}\{(\mathbf{q}^\dagger(n\tfrac{1}{2}l)\mathbf{q}(n\tfrac{1}{2}l))^{(00)} + (\mathbf{q}(n\tfrac{1}{2}l)\mathbf{q}^\dagger(n\tfrac{1}{2}l))^{(00)}\}.$$

The vector $\mathbf{Q}$ is called the quasispin, and this rotation group is denoted as $R_Q(3)$. Not only is $R_Q(3)$ a subgroup of $R(8l + 4)$, but also

$$[\mathbf{Q}, \mathbf{R}^{(\kappa k)}] = 0 \qquad (\kappa + k \text{ odd}).$$

We can, therefore, say that a subgroup of $R(8l + 4)$ is the direct product group $R_Q(3) \times Sp(4l + 2)$.

By restricting the set of $\mathbf{R}^{(\kappa k)}$ with $\kappa + k$ odd to these operators with $\kappa = 0$, k odd, we obtain the generators of the group $R(2l + 1)$. As can easily be seen by reference to Eq. (III-25), these operators commute with the generators of the $R(3)$ group in spin space ($R_S(3)$), which are the operators $\mathbf{R}^{(10)}$. A subgroup of $Sp(4l+2)$ is then $R_s(3) \times R(2l+1)$.

In general, the next useful subgroup of $R(2l+1)$ is $R_L(3)$, the rotation group in the orbital space having the generators $\mathbf{R}^{(01)}$. However, for f electrons, the unexpected vanishing of the 6-j symbol $\{^5_3\,^5_3\,^3_3\}$ allows us to form a group with the operators $\mathbf{R}^{(05)}$ and $\mathbf{R}^{(01)}$. This group does not fall into any of the four general classes of groups, and corresponds to the special group called G_2. Finally, we can form a group from the generators $\mathbf{R}^{(01)} + \mathbf{R}^{(10)}$, which is, of course, another $R(3)$ group, called in this case $R_J(3)$.

5. STATES

We have seen that products of creation and annihilation operators for relativistic (sl) electrons can be taken as the generators of a group $R(8l+5)$ and several of its subgroups. This series of groups and subgroups is written in general as

$$R(8l+5) \supset R(8l+4) \supset R_Q(3) \times Sp(4l+2) \supset R_Q(3)$$
$$\times R_S(3) \times R(2l+1) \supset R_Q(3) \times R_S(3) \times R_L(3) \supset R_Q(3) \times R_J(3). \quad \text{(III-27)}$$

(It should not be assumed that this is the only possible set of subgroups which can be obtained by considering the generators of $R(8l+5)$; see, for example, Judd,[58] and Armstrong and Judd.[59]) Utilization of this knowledge depends in large part on the concept of basis functions.

A set of functions A_k are said to form a *basis* for a group G if any generator of G operating on one of the functions A_k produces another function (or sum of functions) from the same set. Because of the properties of the generators H_i, linear combinations ψ_α of the basis functions A_k can be found, such that

$$H_i \psi_\alpha = m_i \psi_\alpha$$

for all i. The set of numbers m_i obtained by letting H_i run over all q possible values is called the *weight* of ψ_α. The weight of the state ψ_α is said to be higher than that of the state ψ'_α if the first non-vanishing term in the series $m_1 - m'_1$, $m_2 - m'_2, \ldots$ is positive.

The general properties of groups were described at the beginning of Section 4 in terms of an *abstract* group; that is, the elements AB, and so on, were given in no specific form. A *representation* of a group is composed of a set of linear transformations homomorphic to these abstract group elements. An example of a representation of a group is a set of matrices which preserve the group multiplication properties. A matrix representation is said to be *irreducible* if it is impossible to find a similarity transformation which reduces the matrix to block form. Equivalent definitions also apply to algebras. A representation of an algebra can be constructed by describing in matrix form the action of the group generators on a set of basis states ψ_α. It can be shown[40, 53] that an irreducible representation of an algebra formed in the manner just described is completely defined if the weight of the state of highest weight among the basis states is given. Thus, we find it convenient to label basis states not with their own weight, but with the highest weight in the set. Because of the relationship of Lie groups to Lie algebras, the representations of the group is determined if the representation of the algebra is known.

With the generators of the groups of interest given in terms of creation and annihilation operators for (sl) electrons (Section III-4), it is obvious that bases for representations of these algebras can be found among the many electron states of the type

$$\psi = q^\dagger(\beta)q^\dagger(\gamma)\cdots|0\rangle$$

where $\beta = (n\tfrac{1}{2}lm_s m_l)$, $\gamma = (n\tfrac{1}{2}lm'_s m'_l)$, etc. Consider first the group $R(8l + 5)$. The operators H_α can be rewritten in the form

$$H_\alpha = q^\dagger(\alpha)q(\alpha) - \tfrac{1}{2}.$$

H_α operating on the state ψ will have an eigenvalue of $\tfrac{1}{2}$ or $-\tfrac{1}{2}$ according to whether the state (α) is contained in the set of states (β), (γ), etc. which define ψ. Thus the weight of ψ will be of the form $(\pm\tfrac{1}{2}\pm\tfrac{1}{2}\cdots\pm\tfrac{1}{2})$. It is clear that, given a state ψ, and using the operators E_α of (III-24), one can reach any other allowed state ψ' in the shell. Thus, by our definition of a basis, we must use all of the allowed states in a shell to form a basis for $R(8l + 5)$. Further, the highest weight must be the weight with all eigenvalues of the H_α equal to $\tfrac{1}{2}$; this is obtained when ψ is the state of a completely filled shell. We find, then, that all of the states in the configuration

$$(nl)^{N_l}(0 \leq N_l \leq 4l + 2)$$

form the basis for the single representation $(\tfrac{1}{2}\tfrac{1}{2}\cdots\tfrac{1}{2})$ of $R(8l + 5)$.

The basis states for $R(8l + 4)$ can be found in the same manner. The only difference between the two groups is that the generators of $R(8l + 4)$ cannot change the number of electrons in a basis state by an odd number. Thus if ψ contains an even (odd) number of electrons, all operations of the group will result in states containing an even (odd) number of electrons. The states of the shell form, therefore, bases for two distinct representations of $R(8l + 4)$. The H_α operators of $R(8l + 4)$ are identical to those of $R(8l + 5)$; it is easy to see that the highest weight (and therefore the representation) for states of even N is $(\tfrac{1}{2}\tfrac{1}{2}\cdots\tfrac{1}{2})$ and that the highest weight for states of odd N is $(\tfrac{1}{2}\tfrac{1}{2}\cdots\tfrac{1}{2} - \tfrac{1}{2})$.

36   Atomic Structure

Bases for the remaining groups in the chain (III-27) can be found in the same manner. Rather than continue with specific calculations, we list in Tables III-1 and III-2 all possible states of d and f electrons for $N \leq 2l + 1$, along with labels which identify the representation of the groups for which they form bases. In Tables III-1 and -2, the symbol W stands for the representation of the group $R(2l + 1)$; in Table III-2, U stands for the representation of the group G_2. The representations of the groups $R_S(3)$ and $R_L(3)$ can be described simply by the S and L values of the states. The representation of the group $R_Q(3)$ is also described by a single number, Q, called the quasispin of the state. Allowed representations of $Sp(4l + 2)$ are all of the type $(11 \cdots 100 \cdots 0)$; such representations can be described by a single integer v, called the seniority number, which specifies the number of 1's in the representation. It should be noted from Tables III-1 and III-2 that for each seniority there is a unique quasispin; in fact, quasispin and seniority can be shown[43] to be simply related through the equation

$$Q = \tfrac{1}{2}(2l + 1 - v). \tag{III-28}$$

It can also be shown[43] that the z projection of quasispin, M_Q, is related to the number of electrons in the state, n, by the equation

$$M_Q = -\tfrac{1}{2}(2l + 1 - n). \tag{III-29}$$

Table III-1. States of the Configurations d^N

N	v	Q	M_Q	W	S	L
1	1	2	-2	(10)	$\tfrac{1}{2}$	2
2	0	$\tfrac{5}{2}$	$-\tfrac{3}{2}$	(00)	0	0
	2	$\tfrac{3}{2}$	$-\tfrac{3}{2}$	(11)	1	1, 3
				(20)	0	2, 4
3	1	2	-1	(10)	$\tfrac{1}{2}$	2
	3	1	-1	(11)	$\tfrac{3}{2}$	1, 3
				(21)	$\tfrac{1}{2}$	1, 2, 3, 4, 5
4	0	$\tfrac{5}{2}$	$-\tfrac{1}{2}$	(00)	0	0
	2	$\tfrac{3}{2}$	$-\tfrac{1}{2}$	(11)	1	1, 3
				(20)	0	2, 4
	4	$\tfrac{1}{2}$	$-\tfrac{1}{2}$	(10)	2	2
				(21)	1	1, 2, 3, 4, 5
				(22)	0	0, 2, 3, 4, 6
5	1	2	0	(10)	$\tfrac{1}{2}$	2
	3	1	0	(11)	$\tfrac{3}{2}$	1, 3
				(21)	$\tfrac{1}{2}$	1, 2, 3, 4, 5
	5	0	0	(00)	$\tfrac{5}{2}$	0
				(20)	$\tfrac{3}{2}$	2, 4
				(22)	$\tfrac{1}{2}$	0, 2, 3, 4, 6

Table III-2. States of the Configurations f^N

N	v	Q	M_Q	W	U	S	L
1	1	3	-3	(100)	(10)	$\frac{1}{2}$	3
2	0	$\frac{7}{2}$	$-\frac{5}{2}$	(000)	(00)	0	0
	2	$\frac{5}{2}$	$-\frac{5}{2}$	(110)	(10)	1	3
					(11)	1	1, 5
				(200)	(20)	0	2, 4, 6
3	1	3	-2	(100)	(10)	$\frac{1}{2}$	3
	3	2	-2	(111)	(00)	$\frac{3}{2}$	0
					(10)	$\frac{3}{2}$	3
					(20)	$\frac{3}{2}$	2, 4, 6
				(210)	(11)	$\frac{1}{2}$	1, 5
					(20)	$\frac{1}{2}$	2, 4, 6
					(21)	$\frac{1}{2}$	2, 3, 4, 5, 7, 8
4	0	$\frac{7}{2}$	$-\frac{3}{2}$	(000)	(00)	0	0
	2	$\frac{5}{2}$	$-\frac{3}{2}$	(110)	(10)	1	3
					(11)	1	1, 5
				(200)	(20)	0	2, 4, 6
	4	$\frac{3}{2}$	$-\frac{3}{2}$	(111)	(00)	2	0
					(10)	2	3
					(20)	2	2, 4, 6
				(211)	(10)	1	3
					(11)	1	1, 5
					(20)	1	2, 4, 6
					(21)	1	2, 3, 4, 5, 7, 8
					(30)	1	1, 3, 4, 5, 6, 7, 9
				(220)	(20)	0	2, 4, 6
					(21)	0	2, 3, 4, 5, 7, 8
					(22)	0	0, 2, 4, 5, 6, 8, 10
5	1	3	-1	(100)	(10)	$\frac{1}{2}$	3
	3	2	-1	(111)	(00)	$\frac{3}{2}$	0
					(10)	$\frac{3}{2}$	3
					(20)	$\frac{3}{2}$	2, 4, 6
				(210)	(11)	$\frac{1}{2}$	1, 5
					(20)	$\frac{1}{2}$	2, 4, 6
					(21)	$\frac{1}{2}$	2, 3, 4, 5, 7, 8
	5	1	-1	(110)	(10)	$\frac{5}{2}$	3
					(11)	$\frac{5}{2}$	1, 5
				(211)	(10)	$\frac{3}{2}$	3
					(11)	$\frac{3}{2}$	1, 5
					(20)	$\frac{3}{2}$	2, 4, 6
					(21)	$\frac{3}{2}$	2, 3, 4, 5, 7, 8
					(30)	$\frac{3}{2}$	1, 3, 4, 5, 6, 7, 9
				(221)	(10)	$\frac{1}{2}$	3
					(11)	$\frac{1}{2}$	1, 5
					(20)	$\frac{1}{2}$	2, 4, 6
					(21)	$\frac{1}{2}$	2, 3, 4, 5, 7, 8

Table III-2 (*Continued*)

N	v	Q	M_Q	W	U	S	L
					(30)	$\frac{1}{2}$	1, 3, 4, 5, 6, 7, 9
					(31)	$\frac{1}{2}$	1, 2, 3, 3, 4, 5, 5, 6, 6, 7, 7, 8, 9, 10, 11
6	0	$\frac{7}{2}$	$-\frac{1}{2}$	(000)	(00)	0	0
	2	$\frac{5}{2}$	$-\frac{1}{2}$	(110)	(10)	1	3
					(11)	1	1, 5
				(200)	(20)	0	2, 4, 6
	4	$\frac{3}{2}$	$-\frac{1}{2}$	(111)	(00)	2	0
					(10)	2	3
					(20)	2	2, 4, 6
				(211)	(10)	1	3
					(11)	1	1, 5
					(20)	1	2, 4, 6
					(21)	1	2, 3, 4, 5, 7, 8
					(30)	1	1, 3, 4, 5, 6, 7, 9
				(220)	(20)	0	2, 4, 6
					(21)	0	2, 3, 4, 5, 7, 8
					(22)	0	0, 2, 4, 5, 6, 8, 10
	6	$\frac{1}{2}$	$-\frac{1}{2}$	(100)	(10)	3	3
				(210)	(11)	2	1, 5
					(20)	2	2, 4, 6
					(21)	2	2, 3, 4, 5, 7, 8
				(221)	(10)	1	3
					(11)	1	1, 5
					(20)	1	2, 4, 6
					(21)	1	2, 3, 4, 5, 7, 8
					(30)	1	1, 3, 4, 5, 6, 7, 9
					(31)	1	1, 2, 3, 3, 4, 5, 5, 6, 6, 7, 7, 8, 9, 10, 11
				(222)	(00)	0	0
					(10)	0	3
					(20)	0	2, 4, 6
					(30)	0	1, 3, 4, 5, 6, 7, 9
					(40)	0	0, 2, 3, 4, 4, 5, 6, 6, 7, 8, 8, 9, 10, 12
7	1	3	0	(100)	(10)	$\frac{1}{2}$	3
	3	2	0	(111)	(00)	$\frac{3}{2}$	0
					(10)	$\frac{3}{2}$	3
					(20)	$\frac{3}{2}$	2, 4, 6
				(210)	(11)	$\frac{1}{2}$	1, 5

Table III-2 (*Continued*)

N	v	Q	M_Q	W	U	S	L
					(20)	$\frac{1}{2}$	2, 4, 6
					(21)	$\frac{1}{2}$	2, 3, 4, 5, 7, 8
5	1	0	(110)	(10)	$\frac{3}{2}$	3
					(11)	$\frac{5}{2}$	1, 5
				(211)	(10)	$\frac{3}{2}$	3
					(11)	$\frac{3}{2}$	1, 5
					(20)	$\frac{3}{2}$	2, 4, 6
					(21)	$\frac{3}{2}$	2, 3, 4, 5, 7, 8
					(30)	$\frac{3}{2}$	1, 3, 4, 5, 6, 7, 9
				(221)	(10)	$\frac{1}{2}$	3
					(11)	$\frac{1}{2}$	1, 5
					(20)	$\frac{1}{2}$	2, 4, 6
					(21)	$\frac{1}{2}$	2, 3, 4, 5, 7, 8
					(30)	$\frac{1}{2}$	1, 3, 4, 5, 6, 7, 9
					(31)	$\frac{1}{2}$	1, 2, 3, 3, 4, 5, 5, 6, 6, 7, 7, 8, 9, 10, 11
7	0	0	(000)	(00)	$\frac{7}{2}$	0
				(200)	(20)	$\frac{5}{2}$	2, 4, 6
				(220)	(20)	$\frac{3}{2}$	2, 4, 6
					(21)	$\frac{3}{2}$	2, 3, 4, 5, 7, 8
					(22)	$\frac{3}{2}$	0, 2, 4, 5, 6, 8, 10
				(222)	(00)	$\frac{1}{2}$	0
					(10)	$\frac{1}{2}$	3
					(20)	$\frac{1}{2}$	2, 4, 6
					(30)	$\frac{1}{2}$	1, 3, 4, 5, 6, 7, 9
					(40)	$\frac{1}{2}$	0, 2, 3, 4, 4, 5, 6, 6, 7, 8, 8, 9, 10, 12

From these results it is obvious that $n \geq v$ if the relation $|M_Q| \leq Q$ is to hold. It is also clear that one need not label a state by all of the numbers Q, M_Q, v and n, but only by either Q and M_Q or v and n. The states of configuration $l^{4l+2-N}(N \leq 2l+1)$ form bases for the same representations as do the states of l^N with one exception: in keeping with Eq. (III-29), M_Q must be changed to $-M_Q$.

We note that with the exception of pairs of states in f^5, f^6, f^7, f^8, and f^9, all possible states are uniquely defined by the group labels alone. We shall assume in all subsequent discussions that the many-body wavefunctions used form bases for representations of the groups (III-27).

6. PERTURBING INTERACTIONS

Having discussed the techniques for obtaining and classifying the zeroth order wavefunctions, we can now proceed to calculate the effects of the perturbation H_1 (Eq. (III-4c)). Thus far, all electrons of the type (nlj) have the same energy, as do all electrons of the type $(n\frac{1}{2}l)$; H_1 will destroy this degeneracy.

We discuss the interactions by using the second quantized form of the operators. Thus we write for the operator $-\sum_i U(r_i)$:

$$-\sum_i U(r_i) = -\sum q^\dagger(nljm)(nljm|U(r)|n'l'j'm')q(n'l'j'm')$$

where the sum is over n, n', l, l', j, j', m, and m'. A central potential such as $U(r)$ must be diagonal in all quantum numbers except the principal quantum number, n. Thus

$$(nljm|U(r)|n'l'j'm') = \delta(l,l')\,\delta(j,j')\,\delta(m,m')\int (F_{nlj}F_{n'l'j'} + G_{nlj}G_{n'l'j'})U(r)\,dr$$

and the sum above can be written simply as

$$-\sum q^\dagger(nljm)q(n'ljm)\int (F_{nlj}F_{n'lj} + G_{nlj}G_{n'lj})U(r)\,dr$$

where the sum is now over n, n', l, j, and m. Because of the tensorial nature of $q^\dagger(nljm)$ and $q(nljm)$, we can write

$$\sum_m q^\dagger(nljm)q(n'ljm) = -[j]^{1/2}(\mathbf{q}^\dagger(nlj)\mathbf{q}(n'lj))^{(0)}$$

where the two tensors of rank j have been coupled to a tensor of rank 0 on the right above. Then

$$-\sum_i U(r_i) = \sum [j]^{1/2}\int (F_{nlj}F_{n'lj} + G_{nlj}G_{n'lj})U(r)\,dr(\mathbf{q}^\dagger(nlj)\mathbf{q}(n'lj))^{(0)}$$

where the sum on the right is over n, n', l, and j.

This is the form of the perturbation to be used if the zeroth-order many-electron wavefunction is defined in terms of products of creation operators $q^\dagger(nljm)$. If, on the other hand, the zeroth order wavefunction is defined in terms of the operators $q^\dagger(n\frac{1}{2}lm_sm_l)$, the above form for the perturbation is not the most useful; the most useful form in the latter case would be one in which creation and annihilation operators of the type $q^\dagger(n\frac{1}{2}lm_sm_l)$ appear. Because of the definition of this type of creation operator (Eq. III-17), the expression above for $-\sum_i U(r_i)$ can be converted to the desired form simply by using a recoupling coefficient which transforms from jj to LS coupling.[40] Thus

$$-\sum_i U(r_i) = \sum [K][j]^{3/2} \begin{Bmatrix} \frac{1}{2} & l & j \\ \frac{1}{2} & l & j \\ K & K & 0 \end{Bmatrix} \int (F_{nlj} F_{n'lj} + G_{nlj} G_{n'lj}) U(r)\, dr$$

$$\times (\mathbf{q}^\dagger(n\tfrac{1}{2}l)\mathbf{q}(n'\tfrac{1}{2}l))^{(KK)0}_0 \tag{III-30}$$

$$= \sum [K]^{1/2}[j](-1)^{l+j+K-1/2} \begin{Bmatrix} \frac{1}{2} & l & j \\ l & \frac{1}{2} & K \end{Bmatrix}$$

$$\times \int (F_{nlj} F_{n'lj} + G_{nlj} G_{n'lj}) U(r)\, dr\, R^{(KK)0}_0(nl, n'l)$$

where the sum is over n, n', l, j, and K. In the first step on the right above, both the spin and orbital ranks of the tensors $\mathbf{q}^\dagger(n\tfrac{1}{2}l)$ and $\mathbf{q}(n'\tfrac{1}{2}l)$ are coupled to tensors of rank K, and these two tensors are then coupled to a tensor of rank 0.

The two-body interactions appearing in Eq. (III-4c) can be handled in much the same manner as the one-body interaction. First, however, it is convenient to express the two-body interactions in the form

$$\sum_{\substack{k \\ i<j}} \mathbf{g}_i^{(k)} \cdot \mathbf{g}_j^{(k)},$$

where $\mathbf{g}_i^{(k)}$ is a tensor operator of rank k acting only on electron i, etc. This is easily done for the first two-body term in Eq. (III-3c)

$$\sum_{i<j} \frac{e^2}{r_{ij}} = e^2 \sum_{i<j,k} \frac{r_<^k}{r_>^{k+1}} \mathbf{C}_i^{(k)} \cdot \mathbf{C}_j^{(k)} \tag{III-31}$$

where $r_<$ is the smaller of r_i and r_j, $r_>$, the larger, and $\mathbf{r}_{ij} = \mathbf{r}_i - \mathbf{r}_j$, $r_{ij} = |\mathbf{r}_i - \mathbf{r}_j|$. The next term in H_1 requires a recoupling in order to be put into the desired form

$$-\tfrac{1}{2}e^2 \sum_{i<j} \frac{\boldsymbol{\alpha}_i \cdot \boldsymbol{\alpha}_j}{r_{ij}} = \frac{e^2}{2} \sum_{\substack{i<j \\ \kappa}} (\boldsymbol{\alpha}_i \cdot \boldsymbol{\alpha}_j)(\mathbf{C}_i^{(\kappa)} \cdot \mathbf{C}_j^{(\kappa)}) \frac{r_<^\kappa}{r_>^{\kappa+1}}$$

$$= -\frac{e^2}{2} \sum_{i<j,\kappa,k} (\boldsymbol{\alpha}_i\, \mathbf{C}_i^{(\kappa)})^{(k)} \cdot (\boldsymbol{\alpha}_j\, \mathbf{C}_j^{(\kappa)})^{(k)} (-1)^{1+\kappa+k} \frac{r_<^\kappa}{r_>^{\kappa+1}}.$$

$$\tag{III-32}$$

The last two-body term in H_1 is considerably more difficult to handle; we first carry out a recoupling

$$-\frac{e^2}{2} \sum_{i<j} \frac{(\boldsymbol{\alpha}_i \cdot \mathbf{r}_{ij})(\boldsymbol{\alpha}_j \cdot \mathbf{r}_{ij})}{r_{ij}^3} = \sum_{i<j} \left\{ -\tfrac{1}{6}e^2 \frac{\boldsymbol{\alpha}_i \cdot \boldsymbol{\alpha}_j}{r_{ij}} - \frac{(5)^{1/2}}{2} e^2 \frac{((\boldsymbol{\alpha}_i \boldsymbol{\alpha}_j)^{(2)}(\mathbf{r}_{ij}\mathbf{r}_{ij})^{(2)})^{(0)}}{r_{ij}^3} \right\}.$$

42 Atomic Structure

The first term on the right above has the same form as Eq. (III-32) and can simply be added to that term. To handle the second term, we must first obtain an expansion of $(\mathbf{r}_{ij}\mathbf{r}_{ij})^{(2)}/r_{ij}^{3}$. This can be done in a straightforward manner by noting that

$$\nabla_i \frac{1}{r_{ij}} = -\frac{\mathbf{r}_{ij}}{r_{ij}^3}$$

and that $\mathbf{r}_{ij} = r_i \mathbf{C}_i^{(1)} - r_j \mathbf{C}_j^{(1)}$. Then

$$-\frac{\mathbf{r}_{ij}}{r_{ij}^3} = \nabla_i \sum_k \frac{r_<^k}{r_>^{k+1}} \mathbf{C}_i^{(k)} \cdot \mathbf{C}_j^{(k)}$$

$$= \sum_k \nabla_i \frac{r_i^k}{r_j^{k+1}} \mathbf{C}_i^{(k)} \cdot \mathbf{C}_j^{(k)} + \sum_k \nabla_i \frac{r_j^k}{r_i^{k+1}} \mathbf{C}_i^{(k)} \cdot \mathbf{C}_j^{(k)}$$

where the first term is taken if $r_i < r_j$, the second if $r_j < r_i$. When the derivative is taken, this expression becomes

$$\sum_k (-1)^k \left[\frac{k(2k+1)(2k-1)}{3} \right]^{1/2} \frac{r_i^{k-1}}{r_j^{k+1}} (\mathbf{C}_i^{(k-1)}\mathbf{C}_j^{(k)})^{(1)}$$

$$+ \sum_k (-1)^{k+1} \left[\frac{(k+1)(2k+1)(2k+3)}{3} \right]^{1/2} \frac{r_j^k}{r_i^{k+2}} (\mathbf{C}_i^{(k+1)}\mathbf{C}_j^{(k)})^{(1)}.$$

Combining this result with the expression above for $\mathbf{r}_{ij}$ leads immediately to

$$\frac{(\mathbf{r}_{ij}\mathbf{r}_{ij})^{(2)}}{r_{ij}^3} = \sum_k (-1)^k \frac{r_<^k}{r_>^{k+1}} \left\{ (\mathbf{C}_<^{(k)}\mathbf{C}_>^{(k)})^{(2)} \left[\frac{(8k)(k+1)(2k+1)}{(15)(2k-1)(2k+3)} \right]^{1/2} \right.$$

$$- (\mathbf{C}_<^{(k-2)}\mathbf{C}_>^{(k)})^{(2)} \left[\frac{(k)(k-1)(2k-3)(2k+1)}{5(2k-1)} \right]^{1/2}$$

$$\left. + (\mathbf{C}_<^{(k)}\mathbf{C}_>^{(k+2)})^{(2)} \left[\frac{(k)(k+2)(2k+1)(2k+5)}{5(2k+3)} \right]^{1/2} \right\}$$

where $\mathbf{C}_<^{(k)}$ implies $\mathbf{C}_i^{(k)}$ or $\mathbf{C}_j^{(k)}$ as $r_<$ implies r_i or r_j. For simplicity we write this expansion as

$$\frac{(\mathbf{r}_{ij}\mathbf{r}_{ij})^{(2)}}{r_{ij}^3} = \sum_{\kappa, K, k} (-1)^k \frac{r_<^k}{r_>^{k+1}} (\mathbf{C}_<^{(\kappa)}\mathbf{C}_>^{(K)})^{(2)} F(k, \kappa, K)$$

where κ is summed over the values $k, k-2$; K over the values $k, k+2$. Then

$$-\frac{(5)^{1/2}}{2} e^2 \frac{((\alpha_i \alpha_j)^{(2)}(\mathbf{r}_{ij}\mathbf{r}_{ij})^{(2)})^{(0)}}{r_{ij}^3} = \tfrac{5}{2} e^2 \sum \begin{Bmatrix} 1 & 1 & 2 \\ K & \kappa & k \end{Bmatrix}$$

$$\times (\boldsymbol{\alpha}_< \mathbf{C}_<^{(\kappa)})^{(k)} \cdot (\boldsymbol{\alpha}_> \mathbf{C}_>^{(K)})^{(k)} \frac{r_<^{K'}}{r_>^{K'+1}} F(K', \kappa, K) \quad \text{(III-33)}$$

Perturbing Interactions 43

where the sum is over all k and K'; $\kappa = K', K' - 2$; and $K = K', K' + 2$.

Having put all of the two-body interactions into the desired form, we proceed to express them in terms of the second quantization formalism. We write the general term

$$\sum_{k,i<j} \mathbf{g}_i^{(k)} \cdot \mathbf{g}_j^{(k)} = \tfrac{1}{2} \sum q^\dagger(n_1 l_1 j_1 m_1) q^\dagger(n_2 l_2 j_2 m_2)$$
$$\times ([n_1 l_1 j_1 m_1]_1 [n_2 l_2 j_2 m_2]_2 | \mathbf{g}_1^{(k)} \cdot \mathbf{g}_2^{(k)} |$$
$$\times [n_3 l_3 j_3 m_3]_1 [n_4 l_4 j_4 m_4]_2)$$
$$\times q(n_4 l_4 j_4 m_4) q(n_3 l_3 j_3 m_3)$$
$$= \tfrac{1}{2} \sum [k]^{-1} (n_1 l_1 j_1 \| g^{(k)} \| n_3 l_3 j_3)(n_2 l_2 j_2 \| g^{(k)} \| n_4 l_4 j_4)$$
$$\times \{(\mathbf{q}^\dagger(n_1 l_1 j_1) \mathbf{q}(n_3 l_3 j_3))^{(k)} \cdot (\mathbf{q}^\dagger(n_2 l_2 j_2) \mathbf{q}(n_4 l_4 j_4))^{(k)}$$
$$- (\mathbf{q}^\dagger(n_1 l_1 j_1) \mathbf{q}(n_4 l_4 j_4))^{(0)} (-1)^{j_1+j_3} [k][j_1]^{-\tfrac{1}{2}}$$
$$\times \delta(n_2, n_3) \delta(l_2, l_3) \delta(j_2, j_3) \}$$

where the sums are over all values of k, l and j.

This is the form of the interaction which can be used when the wavefunction is defined by a product of creation operators of the type $q^\dagger(nljm)$. The form most useful when the wavefunction is defined by products of operators of the type $q^\dagger(n\tfrac{1}{2}lm_s m_l)$ can be obtained by performing a simple recoupling

$$\sum_{k,i<j} \mathbf{g}_i^{(k)} \cdot \mathbf{g}_j^{(k)} = \sum [k]^{-1} (n_1 l_1 j_1 \| g^{(k)} \| n_3 l_3 j_3)(n_2 l_2 j_2 \| g^{(k)} \| n_4 l_4 j_4)$$
$$\times [k_1, K_1, k_2, K_2, j_1, j_2, j_3, j_4]^{\tfrac{1}{2}} \begin{Bmatrix} \tfrac{1}{2} & \tfrac{1}{2} & k_1 \\ l_1 & l_3 & K_1 \\ j_1 & j_3 & k \end{Bmatrix}$$
$$\times \begin{Bmatrix} \tfrac{1}{2} & \tfrac{1}{2} & k_2 \\ l_2 & l_4 & K_2 \\ j_2 & j_4 & k \end{Bmatrix} \sum_{i<j} \mathbf{r}_i^{(k_1 K_1)k}(n_1 l_1, n_3 l_3) \cdot \mathbf{r}_j^{(k_2 K_2)k}(n_2 l_2, n_4 l_4).$$

(III-34)

Before these expressions for the operators can be used, the reduced matrix elements of the $g^{(k)}$ must be obtained. This can be done in a straightforward but tedious manner; we give here only the results. For the term e^2/r_{ij}:

$$(n_1 l_1 j_1 \| g^{(k)} \| n_3 l_3 j_3)(n_2 l_2 j_2 \| g^{(k)} \| n_4 l_4 j_4)$$
$$= e^2 \iint (n_1 l_1 j_1 \| C^{(k)} \| n_3 l_3 j_3)_1 (n_2 l_2 j_2 \| C^{(k)} \| n_4 l_4 j_4)_2 \frac{r_<^k}{r_>^{k+1}} dr_1 dr_2 \quad \text{(III-35)}$$

where

$$(n_1 l_1 j_1 \| C^{(k)} \| n_3 l_3 j_3) = (-1)^{j_1-\tfrac{1}{2}} \begin{pmatrix} j_1 & k & j_3 \\ -\tfrac{1}{2} & 0 & \tfrac{1}{2} \end{pmatrix} [j_1, j_3]^{\tfrac{1}{2}}$$
$$\times (\Delta(l_1 l_3 k) F_1 F_2 + \Delta(l'_1 l'_2 k) G_1 G_3). \quad \text{(III-36)}$$

44 *Atomic Structure*

The function $\Delta(abc)$ is defined as equal to 1 if a, b, and c can form a triangle and $a + b + c$ is even; $\Delta(abc)$ is equal to 0 otherwise. We have abbreviated F_{nlj_1} as F_1, and so on. The value l' is, as in Section II-2, the orbital momentum of the small component of the wavefunction $|nljm\rangle$. For the term $-(2e^2/3)(\boldsymbol{\alpha}_i \cdot \boldsymbol{\alpha}_j/r_{ij})$:

$$(n_1 l_1 j_1 \| g^{(k)} \| n_3 l_3 j_3)(n_2 l_2 j_2 \| g^{(k)} \| n_4 l_4 j_4)$$

$$= -\frac{2}{3} e^2 \sum_{\kappa} \iint (n_1 l_1 j_1 \|(\alpha \mathbf{C}^{(\kappa)})^{(k)}\| n_3 l_3 j_3)_1$$

$$\times (n_2 l_2 j_2 \|(\alpha \mathbf{C}^{(\kappa)})^{(k)}\| n_4 l_4 j_4)_2$$

$$\times (-1)^{1+\kappa+k} \frac{r_>^k}{r_<^{k+1}} dr_1 \, dr_2 \qquad \text{(III-37)}$$

where

$$(n_1 l_1 j_1 \|(\alpha \mathbf{C}^{(\kappa)})^{(k)}\| n_3 l_3 j_3) = i[k_1, j_1, j_3]^{1/2}$$

$$\times \left\{ \sqrt{2}(-1)^{l_1+\kappa+1} \begin{pmatrix} 1 & \kappa & k \\ -1 & 0 & 1 \end{pmatrix} \begin{pmatrix} j_1 & j_3 & k \\ -\frac{1}{2} & -\frac{1}{2} & 1 \end{pmatrix} (\Delta(l_1 l'_3 \kappa)\right.$$

$$\times F_1 G_3 + \Delta(l'_1 l_3 \kappa) G_1 F_3) + (-1)^{j_3+1/2} \begin{pmatrix} 1 & \kappa & k \\ 0 & 0 & 0 \end{pmatrix}$$

$$\left.\times \begin{pmatrix} j_1 & j_3 & k \\ -\frac{1}{2} & -\frac{1}{2} & 0 \end{pmatrix} (\Delta(l_1 l'_3 \kappa) F_1 G_3 - \Delta(l'_1 l_3 \kappa) G_1 F_3) \right\}. \qquad \text{(III-38)}$$

Finally, we have for the term $-\frac{(5)^{1/2}}{2} e^2 ((\boldsymbol{\alpha}_i \boldsymbol{\alpha}_j)^{(2)} (\mathbf{r}_{ij} \mathbf{r}_{ij})^{(2)})^{(0)}/r_{ij}^3$:

$$(n_1 l_1 j_1 \| g^{(k)} \| n_3 l_3 j_3)(n_2 l_2 j_2 \| g^{(k)} \| n_4 l_4 j_4)$$

$$= \tfrac{5}{2} e^2 \sum F(K', \kappa, K) \begin{Bmatrix} 1 & 1 & 2 \\ K & \kappa & k \end{Bmatrix}$$

$$\times \left\{ \int_0^\infty \int_0^{r_2} \frac{r_1^{K'}}{r_2^{K'+1}} (n_1 l_1 j_1 \|(\alpha \mathbf{C}^{(\kappa)})^{(k)}\| n_3 l_3 j_3)_1 (n_2 l_2 j_2 \|(\alpha \mathbf{C}^{(K)})^{(k)}\| n_4 l_4 j_4)_2 \right.$$

$$\times dr_1 \, dr_2 + \int_0^\infty \int_{r_2}^\infty \frac{r_2^{K'}}{r_1^{K'+1}} (n_1 l_1 j_1 \|(\alpha \mathbf{C}^{(K)})^{(k)}\| n_3 l_3 j_3)_1$$

$$\left. \times (n_2 l_2 j_2 \|(\alpha \mathbf{C}^{(\kappa)})^{(k)}\| n_4 l_4 j_4)_2 \, dr_1 \, dr_2 \right\} \qquad \text{(III-39)}$$

where the sum is over all K', and $\kappa = K'$, $K' - 2$; $K = K'$, $K' + 2$. The reduced matrix elements are given by Eq. (III-38).

What we have done above is to express the interactions felt by the relativistic electrons as a product of a radial function and an angular function. Furthermore, the angular dependence is given in terms of spherical tensors. The rationale behind our desire to write the simple-looking operators of Eq.(III-4c) in the rather imposing and unwieldy looking forms given above is threefold. First, as we shall show in the next section, this latter form is necessary if group theory is to be used to simplify calculations. Secondly, the expansion of the operator H_1 in powers of $(v/c)^2$ is very straightforward if we start from the form obtained in this section. Finally, the operator in this form can be viewed either as a relativistic operator to be evaluated between relativistic wavefunctions, or as an effective operator which can be evaluated between non-relativistic waveunfunctions to produce a relativistic result.[24]

The effective-operator approach is obtained if the operators $\mathbf{r}^{(\kappa k)}(n_\alpha l_\alpha, n_\beta l_\beta)$ of Eqs. (III-30) and (III-34) are replaced with the nonrelativistic operators of Feneuille,[60] $\mathbf{w}^{(\kappa k)}(n_\alpha l_\alpha n_\beta l_\beta)$, which are defined by a reduced matrix element between non-relativistic states

$$\langle n_1 \tfrac{1}{2} l_1 \| w^{(\kappa k)}(n_\alpha l_\alpha, n_\beta l_\beta) \| n_2 \tfrac{1}{2} l_2 \rangle = \delta(n_1, n_\alpha)\, \delta(n_2, n_\beta)\, \delta(l_1, l_\alpha)\, \delta(l_2, l_\beta)[\kappa, k]^{1/2}.$$

(III-40)

We have indicated non-relativistic states in Eq. (III-40) by use of an angular bra and ket; this formalism shall be used throughout the book, with a curved bra or ket being used to indicate a relativistic state. These two apparently unrelated views of the operators of this section are exactly equivalent due to the similarity between Eq. (III-21) and Eq. (III-40). Because of this similarity, we obtain a very important result: a matrix element of $\mathbf{r}^{(\kappa k)}(n_\alpha l_\alpha, n_\beta l_\beta)$ between relativistic many-particle wavefunctions which serve as bases for particular representations of a chain of groups (e.g., the chain given by III-27 if $n_\alpha l_\alpha = n_\beta l_\beta$), will be exactly equal to a matrix element of $\mathbf{w}^{(\kappa k)}(n_\alpha l_\alpha, n_\beta l_\beta)$ between non-relativistic many-particle wavefunctions which serve as bases for the same representations of the same chain of groups. The effective operator approach enables one to evaluate the angular part of the interaction in the non-relativistic scheme, and then to multiply this result by a radial integral calculated using relativistic radial functions, the result being a relativistically valid one. The effective operator approach was discussed in several earlier papers[24,61] before this equivalence between the relativistic and effective operator methods was understood.

7. OPERATORS AND REPRESENTATIONS

Operators, as well as states, can be shown to belong to certain representations of algebras. The discussion of the properties of operators parallels closely the discussion in Section III-5 concerning states. We must find a set of operators O_i which is closed under the group operations S_a. That is, the operators O_i form a representation of a group if

$$S_a O_k S_a^{-1} = O_j$$

for all S_a in the group. If S_a is expanded around the identity operator as was done in Section III-4, the above condition can be written in terms of the infinitesimal operators as

$$[X_\alpha, O_k] = \sum_j b_{\alpha j} O_j.$$

The representation to which the operators belong can be identified by taking combinations of the $O_i(O_\alpha)$ which satisfy the equation

$$[H_i, O_\alpha] = m_i O_\alpha. \qquad \text{(III-41)}$$

The representation can then be identified as in the case of the basis states. Actually, we have already used Eq. (III-41) to define spherical tensors (Section II-4), that is, to find the representation in $R(3)$ to which an operator belongs.

We first apply these procedures to the isolated creation and annihilation operators. Starting with the group $U(4l+2)$, we find that the operators $q^\dagger(n\tfrac{1}{2}lm_s m_l)$ for all values of m_s and m_l form a tensor transforming like the representation $[100\cdots]$; these same operators also form a tensor transforming like $(100\cdots)$ of $Sp(4l+2)$. The operators $q^\dagger(n\tfrac{1}{2}lm_s m_l)$ with m_s fixed and all possible m_l's form a tensor transforming like $(100\cdots)$ of $R(2l+1)$ and l of $R_L(3)$. If $l=3$, this set of operators also form a tensor transforming like (10) of G_2. The operators $q^\dagger(n\tfrac{1}{2}lm_s m_l)$ with m_l fixed and all possible m_s's form a tensor of rank $\tfrac{1}{2}$ in $R_S(3)$. The operators $\tilde{q}(n\tfrac{1}{2}lm_s m_l)$ have exactly the same properties in all of these groups except $U(4l+2)$, where they form a tensor transforming like $[00\cdots 0-1]$. Finally, the two operators $q^\dagger(n\tfrac{1}{2}lm_s m_l)$ and $\tilde{q}(n\tfrac{1}{2}lm_s m_l)$ for fixed m_s, m_l form the components of a tensor of rank $\tfrac{1}{2}$ in $R_Q(3)$, with the creation operator having projection $+\tfrac{1}{2}$, the annihilation operator, projection $-\tfrac{1}{2}$.

The operators $\mathbf{R}^{(\kappa k)}(nl, nl)$ can also be classified in the same manner. The results are shown in Table III-3. Group representation for operators of the type $\mathbf{W}^{(\kappa k)}(nl, n'l')$ have been discussed by Feneuille;[60] as shown in the previous section, this discussion must also serve to define the transformation properties of the $\mathbf{R}^{(\kappa k)}(nl, n'l')$.

In general, all operators of physical interest can be expressed in terms of products of the $\mathbf{R}^{(\kappa k)}(nl, n'l')$. The representations which describe an operator composed of a product of an operator transforming according to an irreducible representation Γ_A and an operator transforming according to an irreducible

Table III-3. Irreducible Representations Formed by the Operators $R^{(kK)}(nl, nl)$

Operator	Group	Representation
$R^{(kK)}$ $k+K$ odd	$R_Q(3) \times Sp(4l+2)$	$0 \times (20 \cdots 0)$
$k+K$ even		$1 \times (110 \cdots 0)$
$R^{(kK)}$ K even (projection π of k fixed)	$R(2l+1)$	$(20 \cdots 0)$
K odd (π fixed)		$(110 \cdots 0)$
$R^{(kK)}$ $K = 0, 2, 4$ (π fixed, $l = 3$)	G_2	(20)
$K = 1, 5$ (π fixed, $l = 3$)		(11)
$K = 3$ (π fixed, $l = 3$)		(10)
$R^{(kK)}$	$R_S(3) \times R_L(3)$	$k \times K$

representation Γ_B can be found by taking a *Kronecker product*. This operation is signified by

$$\Gamma_A \times \Gamma_B = \Sigma_S C_S \Gamma_S$$

and the operator in question will have parts transforming according to several different irreducible representations Γ_S. The coefficient C_S specifies the number of times the representation Γ_S appears in the Kronecker product $\Gamma_A \times \Gamma_B$. Values of C_S for representations of the groups $R(7)$ and $G(2)$ have been given by Nutter.[62] As an example, we consider an operator of the type $O = \sum_{k \text{ even}} \mathbf{r}_i^{(0k)}(3, 3) \cdot \mathbf{r}_j^{(0k)}(3, 3)$. Reference to Table III-3 shows that $\mathbf{r}^{(0k)}$ (k even) transforms according to the representation (200) of $R(7)$; O must therefore transform in $R(7)$ according to the representations contained in the Kronecker product (200) × (200). Using the tables of Nutter, we find (200) × (200) = (000) + (110) + (200) + (220). That is, the operator O can be expanded in a sum of four operators each having definite transformation properties with respect to the group $R(7)$:

$$O = O_1(000) + O_2(110) + O_3(200) + O_4(220).$$

Having now the transformation properties of both states and operators, one can obtain selection rules based on these group properties. Consider a matrix element $(\psi_A | O_B | \psi_C)$ where the subscripts designate the representation of the group G according to which the operator and states transform. This matrix element is just a number; it must remain invariant under any operation of the group G. It is therefore said to be a scalar, and to belong to the identity representation, Γ_I, of G. It can be shown, however, that, if Γ_M is unitary, only Kronecker products of the type $\Gamma_M^* \times \Gamma_M$ contain Γ_I; therefore if the matrix element $(\psi_A | O_B | \psi_C)$, which transforms like $\Gamma_A^* \times (\Gamma_B \times \Gamma_C)$, is to transform like Γ_I, then $\Gamma_B \times \Gamma_C$ must contain Γ_A. If $\Gamma_B \times \Gamma_C$ does not contain Γ_A at least once ($C_A \geq 1$), the matrix element cannot transform like Γ_I, and must

48   *Atomic Structure*

therefore vanish. Numerous selection rules can be obtained by arguments of this type. For example, in $R(3)$, this argument leads to the familiar triangular rules on matrix elements: for $\langle L|W^{(0k)}|L'\rangle$ not to be identically zero, we must be able to form a triangle with the numbers L, k, and L'.

8. THE WIGNER-ECKART THEOREM

As powerful and as useful as they are, the selection rules described in the previous section do not provide the greatest reason for using groups to define states and operators. The most powerful application of group theory to atomic structure comes through the use of the Wigner-Eckart Theorem. We shall not attempt to prove the Wigner-Eckart theorem here; the interested reader is referred to the work of Judd[40] from which these results are taken.

Let us assume a state ψ_i which is one of a set of α functions ($i = 1, \ldots, \alpha$) forming a basis for a representation Γ_A of a group G and another state ϕ_k which is one of a set of β functions ($k = 1, \ldots \beta$) forming a basis for the representation Γ_B of G. Then linear combinations of products of the type $\psi_i \phi_j$ can be formed which transform like basis functions of a representation Γ_M, where $C_M \geq 1$ in $\Gamma_A \times \Gamma_B = \sum_M C_M \Gamma_M$. The proper linear combinations are indicated by the equation

$$\theta_l(\gamma \Gamma_M) = \sum_i (\Gamma_A i; \Gamma_B k|\gamma \Gamma_M l)\psi_i \phi_k \qquad \text{(III-42)}$$

where θ_l is one of the ε functions forming a basis for the representation Γ_M. The symbol γ is required to separate equivalent representations when $C_M > 1$. The coefficient $(\Gamma_A i; \Gamma_B k|\gamma \Gamma_M l)$ is called a coupling or generalized Clebsch-Gordan coefficient.

The Wigner-Eckart theorem can be stated in two forms. The first states that a matrix element $(\phi_k|h_j|\psi_i)$, where ψ_i and ϕ_k are as defined above and h_j is the jth component of a tensor transforming according to the representation Γ_C, can be expressed in the following form:

$$(\phi_k|h_j|\psi_i) = \sum_\gamma (\gamma \Gamma_B k|\Gamma_C j; \Gamma_A i)A(\gamma \Gamma_B, \Gamma_C, \Gamma_A) \qquad \text{(III-43a)}$$

where $A(\gamma \Gamma_B, \Gamma_C, \Gamma_A)$ is independent of i, j, and k; the number of terms appearing in the sum over γ is C_B. The second form of the Wigner-Eckart theorem is useful when more than one set of bases exist for a particular representation. An example of such a case would be the two sets of states $|f^4(211)USL\rangle$ and $|f^5(211)USL\rangle$, both of which form bases for the representation (211) of $R(7)$. The second form of the theorem states that if ψ_i^a, ϕ_k^a and h_j^a are a set of functions independent of the ones defined above, which transform like Γ_A, Γ_B, and Γ_C respectively, then

$$(\phi_k|h_j|\psi_i) = \sum_a B_a(\phi_k^a|h_j^a|\psi_i^a) \qquad \text{(III-43b)}$$

where again the expansion constant is independent of i, j, and k and the number of terms in the sum is equal to C_B.

The first form of the theorem is most useful when the coefficients $(\Gamma_A i; \Gamma_B k | \Gamma_M l)$ are known. The only general case in which they are known is for the group $R(3)$ (and the trivial extension to $R(4) = R(3) \times R(3)$), where they are just the usual Clebsch-Gordan coefficients. The second form is more generally useful and can be used for example, to relate matrix elements in complicated configurations to matrix elements in simpler configurations.

9. NONRELATIVISTIC LIMIT OF THE PERTURBING INTERACTIONS

The forms of the relativistic interactions given in Section 6 can easily be expanded in powers of $(v/c)^2$. The angular operators $\mathbf{R}^{(\kappa k)}$ can be replaced directly with the $\mathbf{W}^{(\kappa k)}$. The radial functions can be expanded as in Section III-2: the large component F_{nlj} becomes R_{nl} and G_{nlj} can be related to R_{nl} through Eq. (III-13). The resulting angular integrals will no longer depend on j, and the sums over j can then be carried out.

We shall not give any details of these straightforward but laborious reductions; details can be found in Reference 61. In addition, we shall consider only the reduction of operators which operate on a configuration l^n. Reduction of the complete set of operators has been carried out by Armstrong and Feneuille.[63] We find that to first order in $(v/c)^2$ the relativistic operators can be written as

1. $e^2 \sum_{i<j, k} \langle l \| C^{(k)} \| l \rangle^2 [k]^{-1} F^k \mathbf{w}_i^{(0k)k} \cdot \mathbf{w}_j^{(0k)k}$

where F^k is a radial integral defined by[32]

$$F^k = \iint [R_{nl}(r_1) R_{nl}(r_2)]^2 \frac{r_<^k}{r_>^{k+1}} \, dr_1 \, dr_2 .$$

This is just $\sum_{i<j} e^2/r_{ij}$ expressed in tensor notation.

2. $-\dfrac{\hbar^2}{2m^2 c^2} \sum_i \int \dfrac{1}{r} \dfrac{dU(r)}{dr} [R_{nl}(r)]^2 \, dr \left[\dfrac{l}{2}(l+1)(2l+1)\right]^{1/2} \mathbf{w}_i^{(11)0}$,

which is the familiar spin-orbit interaction expressed in tensor notation (see also Eq. (V-18)).

3. $-16 \sum_{k, i<j} \dfrac{(2k+1)}{(k+2)} \langle l \| C^{(k)} \| l \rangle^2 l(l+1)(2l+1)$

$$\times \begin{Bmatrix} k & k+1 & 1 \\ l & l & l \end{Bmatrix}^2 M^k \mathbf{w}_i^{(0k+1)k+1} \cdot \mathbf{w}_j^{(0k+1)k+1}.$$

50   Atomic Structure

M^k is a radial integral defined by Marvin:[64]

$$M^k = \tfrac{1}{2}\mu_0^2 \iint \frac{r_<^k}{r_>^{k+3}} [R_{nl}(r_1)R_{nl}(r_2)]^2 \, dr_1 \, dr_2.$$

This term arises classically from the magnetic interaction between the orbits of two electrons.

4. $4\sqrt{5} \sum\limits_{k,\,i<j} \left\{ \dfrac{1}{k+2} \quad \dfrac{1}{k} \quad \dfrac{2}{k+1} \right\} [(k+1)(k+2)(2k+3)]^{1/2} \langle l\|C^{(k)}\|l\rangle$

$$\times \langle l\|C^{(k+2)}\|l\rangle M^k \mathbf{w}_i^{(1k+2)k+1} \cdot \mathbf{w}_j^{(1k)k+1},$$

which can be viewed as arising from the magnetic interaction between the spins of two electrons.

5. $2 \sum\limits_{k,\,i<j} [(k+1)(2l+k+2)(2l-k)]^{1/2} [(-1)^{k+1}[k+1]]^{-1/2}$

$$\times \mathbf{w}_i^{(0k+1)k+1} \cdot \mathbf{w}_j^{(1k)k+1} \{ M^{k-1} \langle l\|C^{(k+1)}\|l\rangle^2 + 2M^k \langle l\|C^{(k)}\|l\rangle^2 \}$$

$$+ (-1)^k [k]^{-1/2} \mathbf{w}_i^{(0k)k} \cdot \mathbf{w}_j^{(1k+1)k} \{ M^k \langle l\|C^{(k)}\|l\rangle^2 + 2M^{k-1} \langle l\|C^{(k+1)}\|l\rangle^2 \}].$$

This describes the magnetic interaction between the spin of one electron and the orbit of another. In the second pair of angular operators appearing in this equation, the term with $k = 0$ must be excluded, as this is part of the spin-orbit interaction (number 2 above).[61] See Sect. V-4a.

6. $\tfrac{4}{3}\mu_0^2 \sum\limits_{k,\beta,\,i<j} (-1)^{k+\beta} \langle l\|C^{(k)}\|l\rangle^2 \int [R_{nl}(r)]^4 \dfrac{1}{r^2} \, dr \, \mathbf{w}_i^{(1k)\beta} \cdot \mathbf{w}_j^{(1k)\beta},$

the interaction arising from the spins of two electrons that are touching.

All terms other than those given above either vanish identically to order $(v/c)^2$ or else have matrix elements which always vanish. The above operators include all possible angular terms from the Breit interaction except the terms $\mathbf{w}_i^{(1k\pm 1)k} \cdot \mathbf{w}_j^{(1k\mp 1)k}$ and $\mathbf{w}_i^{(1k\pm 1)k} \cdot \mathbf{w}_j^{(1k\pm 1)k}$ with k even.

In keeping with the discussion of the previous section, all of these operators have been split up into a sum of terms $\sum a_i A^i$, where A^i is a function of radial integrals only, and a_i is a function of angular operators which has specific transformation properties with respect to the groups of interest. The transformation properties of the possible a_i's are tabulated in Tables III-4 and III-5 for d and f electrons, respectively. It is to be remembered that these separations are based on the angular properties of the operators, and not on the radial properties. Because the $\mathbf{w}^{(\kappa k)}$'s and the $\mathbf{r}^{(\kappa k)}$'s have the same transformation properties, the separation given in these tables applies not only to the non-relativistic operators of this section, but also to the relativistic operators having the same angular dependence.

Table III-4. Symmetry Operators for d-Electron Interactions

Interaction[a]	Coulomb[b]	Spin-Orbit[b]	Spin-Spin[c]	Spin-Other-Orbit[c]	Spin-Spin Contact[d]	Orbit-Orbit
a_1	$^1(00000)(00)^1S_0$	$^1(11000)(11)^3P_0$	$^1(22000)(20)^5D_0$	$^1(22000)(31)^3P_0$	$^f(00)^1S_0$	$^1(00000)(00)^1S_0$
a_2	$^1(22000)(00)^1S_0$		$^1(22000)(22)^5D_0$	$^5(11110)(21)^3P_0$		$^1(22000)(00)^1S_0$
a_3	$^e(22)^1S_0$			$^1(22000)(21)^3P_0$		$^1(22000)(22)^1S_0$
a_4				$^1(22000)(11)^3P_0$		
a_5				$^1(11000)(11)^3P_0$		
a_6				$^3(11000)(11)^3P_0$		
a_7				$^5(11000)(11)^3P_0$		

[a] Representation labels are given as $^{2Q+1}(Sp(10))(R(5))^{2S+1}L_J$.

[b] Judd.[40] The a_1 and a_2 do not correspond to e_0 and e_1 of Judd; e_1 is a superposition of operators transforming as $^1(00000)$ and $^1(22000)$.

[c] B. R. Judd, *Physica* **33**, 174 (1967).

[d] L. Armstrong, Jr., *Phys. Rev.* **170**, 122 (1968).

[e] This operator which is identical to Judd's e_3, is a mixture of operators transforming as $^5(11110)$ and $^1(22000)$. However, $a_3 + \Omega'$, where $\Omega' = 7(V^{(1)})^2 - 3(V^{(3)})^2$, transforms according to the representation $^5(11110)$.

[f] This operator is a mixture of operators transforming as $^1(00000)$ and $^1(22000)$.

Table III-5. Symmetry Operators for f Electron Interactions

Interaction[a] a_i	Coulomb[b]	Spin-Orbit[c]	Spin-Spin[d]
a_1	$^1(0000000)(000)(00)^1S_0$	$^1(1100000)(110)(11)^3P_0$	$^1(2200000)(200)(20)^5D_0$
a_2	$^1(2200000)(000)(00)^1S_0$		$^1(2200000)(220)(20)^5D_0$
a_3	$^1(2200000)(400)(40)^1S_0$		$^1(2200000)(220)(21)^5D_0$
a_4	$^g(220)(22)^1S_0$		$^1(2200000)(220)(22)^5D_0$
a_5			
a_6			
a_7			
a_8			
a_9			
a_{10}			

Spin-Other-Orbit[e]	Spin-Spin-Contact[f]	Orbit-Orbit
$^5(1111000)(211)(11)^3P_0$	$^h(000)(00)^1S_0$	$^1(0000000)(000)(00)^1S_0$
$^5(1111000)(211)(30)^3P_0$	$^1(2200000)(400)(40)^1S_0$	$^1(2200000)(000)(00)^1S_0$
$^1(2200000)(211)(11)^3P_0$		$^1(2200000)(111)(00)^1S_0$
$^1(2200000)(211)(30)^3P_0$		$^1(2200000)(220)(22)^1S_0$
$^1(2200000)(110)(11)^3P_0$		
$^1(2200000)(310)(30)^3P_0$		
$^1(2200000)(310)(31)^3P_0$		
$^1(1100000)(110)(11)^3P_0$		
$^3(1100000)(110)(11)^3P_0$		
$^5(1100000)(110)(11)^3P_0$		

[a] Representation labels are given as $^{2Q+1}(SP(14))(R(7))(G_2)^{2S+1}L_J$.
[b] Racah.[21] The a_1 and a_2 do not correspond to e_0 and e_1 of Racah; e_1 is a superposition of operator transforming as $^1(0000000)$ and $^1(2200000)$.
[c] Judd.[40]
[d] B. R. Judd and H. T. Wadzinski, *J. Math. Phys.* **8**, 2125 (1967).
[e] P. R. Smith and B. G. Wybourne, *J. Math. Phys.* **9**, 1040 (1968).
B. R. Judd, H. M. Crosswhite, and Hannah Crosswhite, *Phys. Rev.* **169**, 130 (1968).
[f] L. Armstrong, Jr., *Phys. Rev.* **170**, 122 (1968).
[g] This operator, which is identical to Racah's e_4, transforms as $^5(1111000)$ and $^1(2200000)$. However $a_4 + \Omega$, where $\Omega = 11(V^{(1)})^2 - 3(V^{(5)})^2$, transforms according to the representation $^5(1111000)$.
[h] This operator is a mixture of operators transforming as $^1(0000000)$ and $^1(2200000)$.

10. DISCUSSION OF ATOMIC STRUCTURE

We have expressed the relativistic atomic structure Hamiltonian in tensor form for two reasons. First, the tensor form is most useful when matrix elements are to be calculated; second, in this form the similarity between the relativistic and non-relativistic Hamiltonians becomes most obvious. Although the relativistic Hamiltonian is incomplete, it is clear that higher order terms will not change this basic similarity.

Actually, the two reasons given above are almost equivalent; calculations can most easily be carried out when the relativistic Hamiltonian is in tensor form simply because when it is in this form, the very powerful Racah techniques of non-relativistic atomic theory can be brought into play. We have outlined in this chapter many of the main points of the techniques introduced by Racah,[19–21] expressed, however, in terminology apropos to the relativistic calculation. Our purpose in doing this has been to demonstrate as clearly as possible the manner in which all of the techniques used in non-relativistic calculations, for example, groups, second quantization, can be directly taken over into the relativistic problem. A complete discussion of these techniques is beyond the scope of this book; excellent discussions are given by Judd.[40,43]

It is in practice very difficult to obtain accurate radial wavefunctions (either relativistic or non-relativistic) for electrons in a complex atom. For that reason, one generally does not attempt to evaluate *a priori* the perturbation Hamiltonian given in the relativistic case by Eq. (III-4c), or in the non-relativistic case, by the equations of Section III-9. Instead, what is usually done is to fit measured energy levels using the non-relativistic Hamiltonian of Section III-9 by treating the radial integrals F^k, M^k, etc. as variable parameters. This fit is accomplished by taking some set of non-relativistic central field wavefunctions ψ_{nr} (i.e., the wavefunctions are separable into radial and angular parts), and evaluating with these wavefunctions matrix elements of the interactions of Section III-9, leaving, however, the radial integrals as unknown parameters. The resulting matrix is then diagonalized in order to obtain eigenvalues which can be compared with measured energies. Of course, before the matrix can be diagonalized, an estimate must be made for the values of the radial integrals; if the resulting eigenvalues of the matrix do not agree with the measured energies of the atom, a revised estimate must be made and the calculation repeated. Techniques exist for determining how to modify the first set of estimates in order to improve the energy fit the second time through the calculation. By this iterative technique, the parameters multiplying the angular functions $\mathbf{w}_i^{(\kappa k)K} \cdot \mathbf{w}_j^{(\kappa' k')K}$ can be determined from experimental data. In addition, the final set of eigenvectors of the matrix, which will be expressed as a sum of products of a coefficient and a basis state ψ_{nr}, are approximate eigenstates of the atom. Obviously, such a calculation fixes only the angular portion of the atomic eigenstates since the radial parts of the ψ_{nr}'s are never actually used. The radial central-field wavefunctions can, in part, be checked by utilizing them to calculate values of F^k, etc., to compare with the values obtained from the fitting procedure. Some care must be taken before imposing this as a test for correctness, however, because the parameterized radial integrals (those obtained from the fitting procedure) often will include higher order effects.[65]

Let us consider what would happen if, instead of the non-relativistic Hamiltonian of Section III-9, we took the relativistic Hamiltonian of Section III-6; and instead of the set of states ψ_{nr}, a set of relativistic wavefunctions ψ_r having the same transformation properties as the states ψ_{nr}. Let us also, for the moment, exclude from the relativistic Hamiltonian those angular terms which do not have counterparts in the non-relativistic Hamiltonian. If all of the coefficients multiplying a given angular interaction were represented by a single coefficient in both the relativistic and non-relativistic calculations, the matrix set up in the relativistic case would be identical to the one which we have described for the non-relativistic case. The technique which would then be used to evaluate relativistic coefficients and relativistic eigenfunctions would be identical to that described above. Finally, the coefficients obtained from the relativistic calculation would be identical in value to the coefficients obtained in the non-relativistic calculation, the eigenfunctions identical except that a non-relativistic basis state having certain transformation properties would be replaced by a relativistic state having the same transformation properties. The explicit forms of the coefficients which have been treated as variable parameters will, of course, be quite different in the relativistic and non-relativistic cases, but the values obtained for the parameters will be the same.

Thus, instead of viewing the results of the traditional fitting procedure as non-relativistic radial integrals and non-relativistic wavefunctions, we can view them as relativistic radial integrals and relativistic wavefunctions. Of course, as we saw in Section III-9, the non-relativistic operators usually used to fit energy levels do not contain all possible angular terms, so if viewed from the standpoint of a relativistic calculation, the fitting procedure does not employ the full Hamiltonian. However, the missing terms in the Hamiltonian can be expected to have only a small effect, since they vanish to order $(v/c)^2$. The fact that even the Hamiltonian of Section III-6 is not relativistically correct does not influence this parameterization technique at all, since the only result of including higher order relativistic terms would be to change the explicit form of the fitted parameters. Thus, our interpretation of the meaning of the parameters might be slightly altered by the inclusion of higher order terms, but neither the values of the parameters nor the compositions of the eigenfunctions would be affected.

It is interesting to question whether any of these group operations discussed above lead to "good" quantum numbers; that is, whether an eigenfunction of the fine structure Hamiltonian could also be an eigenfunction of any of the commuting generators, the H_i's, of the groups (III-27). This question is easily answered by reference to Tables III-4 and -5 which describe the transformation properties of the main components of the perturbation Hamiltonian. For the Hamiltonian and the commuting group generators to have a common eigenfunction, the Hamiltonian and the generators must themselves commute. From the discussion of Section III-7, we know that an operator which commutes with all of the H_i's in a group must belong to the representation (00 $\cdots$ 0) of that group. As can be seen from Tables III-4 and -5, however, the total Hamiltonian belongs to this scalar representation only in $R_J(3)$. Thus J is the only good quantum number which we can attach to the eigenfunctions of the fine structure Hamiltonian.

Finally, we note that the wavefunction function found by the diagonalization process above, ψ_J, satisfies the eigenvalue equation

$$H_{fs}\psi_J \equiv (H_0 + H_1')\psi_J = E_J\psi_J. \tag{III-44}$$

Here, H_1' is the operator H_1 modified in such a way that it has no non-vanishing matrix elements between any of the states used to form the energy matrix and any state not used in setting up this matrix. How closely Eq.(III-44) approaches the eigenvalue equation

$$H\phi = E\phi$$

where H is the total Hamiltonian of Eq. (III-3) depends entirely on the set of wavefunctions ψ_r used to set up the energy matrix. If a complete set of functions is used (e.g. all bound and free central-field wavefunctions of the atom), the equivalence will be exact. Of course, such a matrix would be too large to diagonalize. If it could be diagonalized, essentially all free and bound state energies of the atom would have to be known in order to fit the large number of variable parameters which would appear. Finally, even if these objections could be overcome, use of a complete set would still be impractical since many of the matrix elements would be too small to be fit properly due to uncertainties in measured energies. In practice, the energy matrix is most often set up using only wavefunctions belonging to a single configuration; the effects of excited configurations are then handled by means of perturbation theory.[65] As we shall see in Chapter V, use of this restricted set to determine ψ_J can have serious effects on the calculated hyperfine structure.

In the remainder of this book, we shall consider perturbations to the zeroth order wavefunctions ψ_J caused by incorrect treatment in H_{fs} of the interaction between the atomic electrons and the nucleus. We shall discuss the differences between H (Eq. III-3) and H_{fs} only insofar as it may effect the interaction between the electrons and the nucleus.

IV. The Hyperfine Interaction

1. ELECTROMAGNETIC FIELDS OF A NONSPHERICAL NUCLEUS

In the preceding chapter we discussed the interactions between electrons moving in a field of a spherical point nucleus of charge Ze, and saw how these interactions lead to the atomic fine structure. In many cases, however, the assumption of a spherical point nucleus is incorrect. The deviation of the nucleus from this ideal form leads to a further interaction between the nucleus and the orbital electrons which is in general much smaller than the fine structure interaction itself. The effects of this additional interaction, called the hyperfine interaction, are small enough to be treated accurately using perturbation theory by taking as unperturbed states the fine structure states of the preceding section.

Just as the interaction between the spherical point nucleus and the atomic electrons is represented by a sum of one electron operators $(\sum_i Ze^2/r_i)$, so may the hyperfine interaction be represented by a sum of one electron operators. We can therefore confine our attention, for the moment, to a study of the interaction between a nonspherical, finite nucleus and the ith electron in an N electron atom. The problem to be considered will be that of a relativistic electron moving in the field of a stationary non-relativistic nucleus. The assumption of a nonrelativistic nucleus is reasonable due to the large mass (931 MeV) and relatively small energy (20–30 MeV) of the nucleons which compose the nucleus. We shall also treat the nucleus classically in the sense that all nuclear electromagnetic fields and currents will be assumed to arise from the protons and neutrons only.

In this approximation the nucleus has a charge density $\rho = \psi_N^*(\sum_n g_{ln} e)\psi_{N'}$ where $g_{ln} = 1$ for protons, 0 for neutrons. We are interested in steady state solutions $(\partial \rho/\partial t = 0)$, so we require that ψ_N and $\psi_{N'}$ be eigenfunctions of a nuclear Hamiltonian corresponding to the same eigenvalue. The Hamiltonian for such a classical nucleus in the absence of an applied external field must commute with $\mathbf{I} = \mathbf{L}_N + \mathbf{S}_N$, where $\mathbf{L}_N$ is the sum of the orbital momenta of the nucleons; $\mathbf{S}_N$, the sum of the spin momenta. (It is possible that the nuclear Hamiltonian might commute with either $\mathbf{L}_N$ or $\mathbf{S}_N$ separately. However, most nuclear models call for a strong spin-orbit interaction, which does not commute with either $\mathbf{L}_N$ or $\mathbf{S}_N$ but only with their sum, $\mathbf{I}$.) In addition, since there is no external field present, there can be no preferred direction in space. This means that there must be a $2I + 1$-fold energy degeneracy corresponding to wavefunctions having a common eigenvalue of $\mathbf{I}^2$, but different eigenvalues of I_z. Thus, in general, we expect ψ_N and $\psi_{N'}$ to differ only in m_I, the eigenvalue of I_z.

The electrostatic potential at the electron position due to this nuclear charge density is easily found; it is just

$$V(r_e) = \sum_n \int \frac{\rho(r_n)}{|\mathbf{r}_e - \mathbf{r}_n|} d\tau_N$$

$$= e \sum_{n,k} \int \psi_N^* \frac{r_<^k}{r_>^{k+1}} \mathbf{C}_e^{(k)} \cdot \mathbf{C}_n^{(k)} g_{ln} \psi_{N'} d\tau_N. \quad \text{(IV-1)}$$

The first term in this series when $r_e > r_n$ is Ze/r_e, the potential for a spherical nucleus; the remainder of the terms in the sum result from deviations from this spherical form.

The interaction between a relativistic electron and a magnetic field is introduced into the Dirac Hamiltonian by replacing $\boldsymbol{\alpha} \cdot c\mathbf{p}$ by $\boldsymbol{\alpha} \cdot (c\mathbf{p} + e\mathbf{A})$, where $\mathbf{A}$ is the vector potential. Thus, to find the interaction between the electrons and the magnetic field of the nucleus, we need to find the vector potential produced by the currents of the nucleus. This current can be expressed as a sum of two parts; a convection current

$$\mathbf{j}_c = \frac{e}{2} \left[\psi_N^* \sum_n g_{ln} \mathbf{v}_n \psi_{N'} - \psi_{N'} \sum_n g_{ln} \mathbf{v}_n \psi_N^* \right] \quad \text{(IV-2a)}$$

and a spin current

$$\mathbf{j}_s = \nabla \times \psi_N^* \sum_n g_{sn} \frac{e\hbar}{2M_p} \mathbf{S}_n \psi_{N'} \quad \text{(IV-2b)}$$

where $g_{sn} = -3.8263$ for neutrons and 5.5856 for protons; M_p = proton mass. The condition that the current be steady state, that is $\nabla \cdot \mathbf{j} = 0$, places the same limitations on ψ_N and $\psi_{N'}$ as found in the discussion of the requirement that $\partial \rho / \partial t = 0$.

Any vector field can be expressed in terms of the vector operators $\nabla \phi_1$, $\mathbf{L}\phi_2$ and $\nabla \times \mathbf{L}\phi_3$, where the ϕ's are scalars.[38] If $\mathbf{A}$ is expanded in a series of this type, the part $\nabla \phi_1$ will just be of the form of a gauge transformation.

However, the Dirac equation is invariant under gauge transformations.[66] This means, for example, that if ψ_i is the solution to the Dirac equation in the absence of a vector potential (Eq. (III-4b)), then $e^{ie\phi_1}\psi_i$ is a solution for a Dirac equation containing a vector potential term of the form $\mathbf{A} = \nabla\phi_1$. The effect of a gauge transformation, then, is simply to multiply the wavefunctions by a phase which is the same for all states of the atom. Since all matrix elements involve a product $\psi_i^*\psi_j$, this change in the wavefunctions cannot affect any calculation. We may therefore neglect any part of $\mathbf{A}$ which is of the form $\nabla\phi$, and can expand $\mathbf{A}$ in terms of the operators $\mathbf{L}\phi_2$ and $\nabla \times \mathbf{L}\phi_3$ only.

Without loss of generality, we may partially choose the forms of ϕ_2 and ϕ_3 such that

$$\mathbf{A}(r_e) = \sum_{k,q} (\mathbf{L}_e C_{eq}^{(k)} \phi_2(kq) + \nabla_e \times \mathbf{L}_e C_{eq}^{(k)} r_e^k \phi_3(kq)) \qquad \text{(IV-3)}$$

where $\phi_2(kq)$ and $\phi_3(kq)$ are scalar functions of r_e and r_n. Explicit forms for $\phi_2(kq)$ and $\phi_3(kq)$ are most easily found if we use the Coulomb gauge, $\nabla \cdot \mathbf{A} = 0$. In this gauge, the vector potential at the electron position is related to currents caused by the nucleons through the equation

$$\mathbf{A}(r_e) = \frac{1}{c} \sum_n \int \frac{\mathbf{j}(r_n)}{|\mathbf{r}_e - \mathbf{r}_n|} d\tau_N \qquad \text{(IV-4)}$$

where the sum is over all nucleons. We solve first for $\phi_2(kq)$; this is done by equating (IV-3) and (IV-4), taking the dot product of both sides with $(\mathbf{L}C_\pi^{(\kappa)})^*$ and integrating over the angular coordinates. Then, utilizing the normalization integral for vector spherical harmonics[38]

$$\int (\mathbf{L} C_\pi^{(\kappa)})^* \cdot \mathbf{L} C_q^{(k)} \, d\Omega = \delta(\kappa, k) \, \delta(\pi, q) \frac{4\pi(k)(k+1)}{2k+1}, \qquad \text{(IV-5)}$$

and the fact that $\mathbf{L}$ is orthogonal to $\nabla \times \mathbf{L}$, we obtain

$$\phi_2(kq) = \frac{(2k+1)}{4\pi k(k+1)c} \sum_n \iint \frac{(\mathbf{L}_e C_e^{(k)})^* \cdot \mathbf{j}(r_n)}{|\mathbf{r}_e - \mathbf{r}_n|} d\tau_N \, d\Omega_e.$$

To simplify the term on the right, we note that $|\mathbf{r}_1 - \mathbf{r}_2|^{-1}$ can be expanded as

$$\frac{1}{|\mathbf{r}_1 - \mathbf{r}_2|} = \sum_{k,q} \frac{r_<^k}{r_>^{k+1}} \frac{1}{k(k+1)} (\mathbf{L}_1 C_{1q}^{(k)})^* \cdot (\mathbf{L}_2 C_{2q}^{(k)}). \qquad \text{(IV-6)}$$

This expansion can easily be shown to be equivalent to the more common expansion (Eq. (III-31)) by explicitly writing out the terms $\mathbf{L}C_q^{(k)}$ and carrying out a sum over 3-j symbols. Using this expansion, and the normalization integral of Eq. (IV-5), we find

$$\phi_2(kq) = \frac{1}{ck(k+1)} \sum_n \int \frac{r_<^k}{r_>^{k+1}} (\mathbf{L}_n C_{nq}^{(k)})^* \cdot \mathbf{j}(r_n) \, d\tau_N. \qquad \text{(IV-7)}$$

The contribution of the convection current to $\phi_2(kq)$ is found by combining Eqs. (IV-2a) and (IV-7):

$$\frac{-i\beta_N}{k(k+1)} \sum_n \int \left(\mathbf{L}_n \frac{r_<^k}{r_>^{k+1}} C_{nq}^{(k)} \right)^* \cdot (\psi_N^* g_{ln} \mathbf{V}_n \psi_{N'} - \psi_{N'} g_{ln} \mathbf{V}_n \psi_N^*) \, d\tau_N$$

$$= \frac{\beta_N}{k(k+1)} \sum_n \int \mathbf{r}_n \times \mathbf{V}_n \left(\frac{r_<^k}{r_>^{k+1}} C_{nq}^{(k)} \right)^* \cdot (\psi_N^* g_{ln} \mathbf{V}_n \psi_{N'} - \psi_{N'} g_{ln} \mathbf{V}_n \psi_N^*) \, d\tau_N$$

$$= \frac{-i\beta_N}{k(k+1)} \sum_n \int \mathbf{V}_n \left(\frac{r_<^k}{r_>^{k+1}} C_{nq}^{(k)} \right)^* \cdot (\psi_N^* g_{ln} \mathbf{L}_n \psi_{N'} - \psi_{N'} g_{ln} \mathbf{L}_n \psi_N^*) \, d\tau_N$$

where $\beta_N = eh/2M_p c$ (M_p = proton mass). Using the hermiticity of $\mathbf{L}_n$ and the vector identity $\mathbf{L} \cdot \mathbf{V} = 0$, the above becomes finally

$$\frac{-2i}{k(k+1)} \int \psi_N^* \sum_n \mathbf{V}_n \left(\frac{r_<^k}{r_>^{k+1}} C_{nq}^{(k)} \right)^* \cdot g_{ln} \beta_N \mathbf{L}_n \psi_{N'} \, d\tau_N \qquad \text{(IV-8)}$$

The contribution to $\phi_2(kq)$ from the spin current is given by

$$\frac{1}{k(k+1)} \sum_n \int \left(\mathbf{L}_n \frac{r_<^k}{r_>^{k+1}} C_{nq}^{(k)} \right)^* \cdot \mathbf{V}_n \times \psi_N^* g_{sn} \beta_N \mathbf{S}_n \psi_{N'} \, d\tau_N$$

$$= \frac{1}{k(k+1)} \sum_n \int \left(\mathbf{V}_n \times \mathbf{L}_n \frac{r_<^k}{r_>^{k+1}} C_{nq}^{(k)} \right)^* \cdot \psi_N^* g_{sn} \beta_N \mathbf{S}_n \psi_{N'} \, d\tau_N. \qquad \text{(IV-9)}$$

Further simplification is possible through use of the identities[38]

$$\mathbf{V} \times \mathbf{L}(r^k C_q^{(k)}) = i(k+1) \mathbf{V} r^k C_q^{(k)}$$

$$\mathbf{V} \times \mathbf{L}(r^{-k-1} C_q^{(k)}) = -ik \mathbf{V}(r^{-k-1} C_q^{(k)}) + 4\pi i \mathbf{r} \, \delta_{kq}.$$

The term δ_{kq} is the kth derivative of a three-dimensional delta function with a singularity at the origin. Scrutiny of Eq. (IV-9) shows, however, that this delta function makes no contribution to $\phi_2(kq)$. The function δ_{kq} appears only when $r_e < r_n$, and the integral over δ_{kq} can be non-vanishing only if it includes the origin. Thus the integral can be non-zero if $r_e = 0$. Since, for $r_e < r_n$, $\phi_2(kq)$ is proportional to r_e^k, we have the further limitation that the integral over the delta function may contribute only to $\phi_2(00)$. The value of the integral

$$\int_0^\infty \delta_{kq} \psi_N^* \mathbf{r}_n \cdot \mathbf{S}_n \psi_N \, d\tau_N$$

is given by the coefficient of $r_n^k C_{nq}^{(k)}$ in a power series expansion of the integrand around the origin. The integrand, however, has negative parity, so the expansion must be made in terms of the $C^{(k)}$ with k odd. Thus, the integral must vanish for $k = 0$, and there can be no contribution from the expression

Electromagnetic Fields

containing the delta function to $\phi_2(00)$ either. The delta function can therefore be completely neglected.

Let us now consider the term containing $\phi_3(kq)$. Eq. (IV-4) can be put into a form which makes the solution for $\phi_3(kq)$ very straightforward:

$$\mathbf{V}_e \times \mathbf{A}(\mathbf{r}_e) = \mathbf{V}_e \times \frac{1}{c}\sum_n \int \frac{\mathbf{j}(\mathbf{r}_n)}{|\mathbf{r}_e - \mathbf{r}_n|}\, d\tau_N$$

$$= -\frac{1}{c}\sum_n \int \frac{\mathbf{V}_n \times \mathbf{j}(\mathbf{r}_n)}{|\mathbf{r}_e - \mathbf{r}_n|}\, d\tau_N$$

or

$$\sum_{k,q}(\mathbf{V}_e \times \mathbf{L}_e\, C_{eq}^{(k)}\phi_2(kq) + \mathbf{V}_e \times (\mathbf{V}_e \times \mathbf{L}_e)C_{eq}^{(k)}r_e^{\,k}\phi_3(kq))$$

$$= -\frac{1}{c}\sum_n \int \frac{\mathbf{V}_n \times \mathbf{j}(\mathbf{r}_n)}{|\mathbf{r}_e - \mathbf{r}_n|}\, d\tau_N.$$

Using $\mathbf{V} \times (\mathbf{V} \times \mathbf{L}) = \mathbf{V}^2\mathbf{L}$ and $\mathbf{V}^2 C_q^{(k)}r^k = 0$, this becomes

$$\sum_{kq}[\mathbf{V}_e \times \mathbf{L}_e\, C_{eq}^{(k)}\phi_2(kq) + \mathbf{L}_e\, C_{eq}^{(k)}r_e^{\,k}(\mathbf{V}_e^2\phi_3(kq))] = -\frac{1}{c}\sum_n \int \frac{\mathbf{V}_n \times \mathbf{j}(\mathbf{r}_n)}{|\mathbf{r}_e - \mathbf{r}_n|}\, d\tau_N. \quad \text{(IV-10)}$$

If we define $\phi_4(kq)$ by the equation $\phi_4(kq) = r_e^{\,k}\mathbf{V}_e^2\phi_3(kq)$, then Eq. (IV-10) becomes very similar to Eqs. (IV-3) and (IV-4) except that $\mathbf{j}(\mathbf{r}_n)$ has been replaced by $\mathbf{V} \times \mathbf{j}(\mathbf{r}_n)$. Obviously, $\phi_4(kq)$ is given by

$$\phi_4(kq) = \frac{1}{ck(k+1)}\sum_n \int \left(\mathbf{L}_n \frac{r_<^k}{r_>^{k+1}} C_{nq}^{(k)}\right)^* \cdot \mathbf{V}_n \times \mathbf{j}(\mathbf{r}_n)\, d\tau_N$$

$$= \frac{1}{ck(k+1)}\sum_n \int \left(\mathbf{V}_n \times \mathbf{L}_n \frac{r_<^k}{r_>^{k+1}} C_{nq}^{(k)}\right)^* \cdot \mathbf{j}(\mathbf{r}_n)\, d\tau_N.$$

Through use of the identities given below Eq. (IV-9) and $\mathbf{V} \cdot \mathbf{j} = 0$, we obtain

$$\mathbf{V}_e^2\phi_3(kq) = \frac{4\pi i}{ck(k+1)}\sum_n \int_{r_e}^{\infty} \delta_{kq}\mathbf{r}_n \cdot \mathbf{j}_n\, d\tau_N$$

which reveals that $\mathbf{V}_e^2\phi_3(kq) = 0$ unless $r_e = 0$, when it takes on some finite value. The consequences of such a condition on $\phi_3(kq)$ are easily seen if we write

$$\mathbf{V}_e \times \mathbf{L}_e(r_e^{\,k}C_{eq}^{(k)}\phi_3(kq)) = i\mathbf{V}_e[\mathbf{V}_e \cdot \mathbf{r}_e\, r_e^{\,k}C_{eq}^{(k)}\phi_3(kq)] - i\mathbf{V}_e^2\mathbf{r}_e\, r_e^{\,k}C_{eq}^{(k)}\phi_3(kq)$$

The only portion of this expression that cannot be written in the form of a gauge transformation is given by

$$\mathbf{r}_e r_e^{\,k} C_{eq}^{(k)} \mathbf{V}_e^2 \phi_3(kq)$$

62 The Hyperfine Interaction

which vanishes identically because of the value of $\mathbf{V}_e^2 \phi_3(kq)$. Thus, the portion of $\mathbf{A}$ written as $\mathbf{V} \times \mathbf{L}\phi_3$ can also be put into the form of a gauge transformation, and can be neglected henceforth.

As a result of the above discussion, we find that we can express $\mathbf{A}$ in the form

$$\mathbf{A}(r_e) = \sum_{k,q} \frac{-i\beta_N}{k(k+1)} \mathbf{L}_e C_{eq}^{(k)}$$

$$\times \left[\frac{1}{r_e^{k+1}} \int_0^{r_e} \psi_N^* \sum_n \mathbf{V}(r_n^k C_{nq}^{(k)})^* \cdot (2g_{ln}\mathbf{L}_n + (k+1)g_{sn}\mathbf{S}_n)\psi_{N'} \, d\tau_N \right.$$

$$\left. + r_e^k \int_{r_e}^\infty \psi_N^* \sum_n \mathbf{V}(r_n^{-k-1} C_{nq}^{(k)})^* \cdot (2g_{ln}\mathbf{L}_n - kg_{sn}\mathbf{S}_n)\psi_{N'} \, d\tau_N \right].$$

(IV-11)

In the remainder of this chapter, we shall assume that $r_e > r_n$ in both the magnetic (Eq. (IV-11)) and electrostatic (Eq. (IV-1)) interactions. The errors associated with such an assumption will be discussed in a later chapter.

2. THE HYPERFINE HAMILTONIAN

The Hamiltonian which describes the perturbation on the fine structure energy levels of the atom due to the non-spherical nature of the nucleus, H_{hyp}, is given by the difference between the potentials obtained in the preceding section and $\sum_i - Ze^2/r_i$:

$$H_{\text{hyp}} = \sum_i \left(\frac{Ze^2}{r_i} - eV(r_i) + \boldsymbol{\alpha}_i \cdot e\mathbf{A}(r_i) \right).$$

(IV-12)

If the expressions $eV(r)$ and $\mathbf{A}(r)$ obtained above are used directly in Eq. (IV-12), H_{hyp} acts only in the space of the electrons. However, it is usually more convenient if we write H_{hyp} in such a form that it is an operator in both the electronic and nuclear spaces. This can easily be done if we assume that in the future, our wavefunctions will be written as a (possibly coupled) product of nuclear and electronic wavefunctions

$$\Psi = \psi_N \psi_e.$$

Then, taking into account the approximation made at the end of the preceding section, we find that we can write

$$H_{\text{hyp}} = \sum_i \sum_{k>0} \left(-\frac{e}{r_i^{k+1}} \mathbf{C}_i^{(k)} \cdot \mathbf{F}^{(k)} - \frac{ie}{k} \boldsymbol{\alpha}_i \cdot \sum_q L C_{iq}^{(k)} \frac{1}{r_i^{k+1}} N_{-q}^{(k)}(-1)^q \right)$$

$$= \sum_i \sum_{k>0} \left(-\frac{e}{r_i^{k+1}} \mathbf{C}_i^{(k)} \cdot \mathbf{F}^{(k)} + \frac{ie}{r_i^{k+1}} \sqrt{\left(\frac{k+1}{k}\right)} (\boldsymbol{\alpha}_i \mathbf{C}_i^{(k)})^{(k)} \cdot \mathbf{N}^{(k)} \right).$$

(IV-13)

$F_q^{(k)}$, the electrostatic nuclear operator, is given by

$$F_q^{(k)} = e \sum_n g_{ln} r_n^k C_{nq}^{(k)} \tag{IV-14}$$

and $N_q^{(k)}$, the magnetic nuclear operator is given by

$$N_q^{(k)} = \beta_N \sum_n (-1)^q \nabla(r_n^k C_{n-q}^{(k)})^* \cdot \left(\frac{2g_{ln}}{k+1} \mathbf{L}_n + g_{sn} \mathbf{S}_n \right)$$

$$= \beta_N \sum_n [k(2k-1)]^{1/2} r_n^{k-1} \left[\frac{2g_{ln}}{k+1} (\mathbf{C}_n^{(k-1)} \mathbf{L}_n)_q^{(k)} + g_{sn} (\mathbf{C}_n^{(k-1)} \mathbf{S}_n)_q^{(k)} \right].$$

(IV-15)

The operators $\mathbf{F}^{(k)}$ transform in the same way as $\mathbf{C}_n^{(k)}$, that is, like a tensor of rank k in the nuclear space. The operators $\mathbf{N}^{(k)}$ also transform like a tensor of rank k in nuclear space, as can be seen in the second form of Eq. (IV-15). It is clear that $\mathbf{F}^{(k)}$ has the parity of $\mathbf{C}^{(k)}$, for example $(-1)^k$, whereas $\mathbf{N}^{(k)}$ has the parity of $\mathbf{C}^{(k-1)}$, or $(-1)^{k-1}$. If we assume that the nuclear wavefunction has well-defined parity, then matrix elements of $\mathbf{F}^{(k)}$ between states of like parity must vanish for k odd; those of $\mathbf{N}^{(k)}$, for k even.

Although the expansion over k in Eq. (IV-13) has no upper limit, effective upper limits are set by both the nuclear and electronic wavefunctions. The kth term in Eq. (IV-13) is composed of an operator transforming like a tensor of rank k in the $R_J(3)$ space of the electrons and an operator transforming like a tensor of rank k in the $R_I(3)$ space of the nucleons. All matrix elements diagonal in the quantum numbers I and J must therefore vanish if either $k > 2I$ or $k > 2J$. This result arises from the familiar triangular conditions on matrix elements, which can be clearly seen in Eq. (II-19).

The hyperfine interaction obviously does not commute with the generators of $R_J(3)$ and $R_I(3)$ and as a result J and I cannot be considered to be good quantum numbers when hyperfine interactions are taken into account. However, the hyperfine interaction is certainly a scalar in the $R_F(3)$ space formed by the generators $\mathbf{F} = \mathbf{I} + \mathbf{J}$. Thus, strictly speaking, F is the only good quantum number when the hyperfine interaction is taken into account. On the other hand, because of the large energy difference between nuclear states of different I (100 keV) and small hyperfine energy differences (kmc), we can make the approximation that I remains a good quantum number even when hyperfine interactions are considered. Not only does this assumption simplify our considerations, but it also is necessary if we are not to violate the steady state condition placed on the nuclear current and charge density in our derivation of the nuclear electrostatic and magnetic fields.

3. THE HYPERFINE HAMILTONIAN FOR THE MANY-ELECTRON ATOM

The hyperfine interaction for the many-electron atom can be most easily handled by considering the second quantized form of the interaction. Thus, using Eq. (IV-13), we find

$$H_{\text{hyp}} = \sum q^{\dagger}(nljm)(nljm|H_{\text{hyp}}(i)|n'l'j'm')q(n'l'j'm')$$
$$= -\sum_k [k]^{-1/2}(\mathbf{q}^{\dagger}(nlj)\mathbf{q}(n'l'j'))^{(k)}$$
$$\left[-e\mathbf{F}^{(k)}(nlj\|r^{-k-1}C^{(k)}\|n'l'j') + ie\sqrt{\left(\frac{k+1}{k}\right)} \right.$$
$$\left. \times \mathbf{N}^{(k)}(nlj\|r^{-k-1}(\alpha C^{(k)})^{(k)}\|n'l'j') \right]. \quad \text{(IV-16)}$$

When the wavefunction is expressed in terms of creation operators of the type $q^{\dagger}(n\tfrac{1}{2}l)$, H_{hyp} should be expressed in terms of the operators $\mathbf{R}^{\kappa k}$. This is easily done using the techniques of Section III-6. Thus

$$H_{\text{hyp}} = \sum \begin{Bmatrix} \tfrac{1}{2} & l & j \\ \tfrac{1}{2} & l' & j' \\ S & L & k \end{Bmatrix} [j, j', S, L]^{1/2}[k]^{-1/2} \mathbf{R}^{(SL)k}(nl, n'l')$$
$$\cdot \left[-e\mathbf{F}^{(k)}(nlj\|r^{-k-1}C^{(k)}\|n'l'j') + ie\sqrt{\left(\frac{k+1}{k}\right)} \right.$$
$$\left. \times \mathbf{N}^{(k)}(nlj\|r^{-k-1}(\alpha C^{(k)})^{(k)}\|n'l'j') \right] \quad \text{(IV-17)}$$

where the sum is over $k > 0$, nl, $n'l'$, j, j', S, and L. This can be written as

$$H_{\text{hyp}} = \sum_{k>0} (\mathbf{Q}^{(k)} \cdot \mathbf{F}^{(k)} + \mathbf{M}^{(k)} \cdot \mathbf{N}^{(k)}) \quad \text{(IV-18)}$$

if we define the tensor operators acting in the space of the electrons

$$\mathbf{Q}^{(k)} = -e \sum \begin{Bmatrix} \tfrac{1}{2} & l & j \\ \tfrac{1}{2} & l' & j' \\ S & L & k \end{Bmatrix} [j, j', S, L]^{1/2}[k]^{-1/2}$$
$$\times (nlj\|r^{-k-1}C^{(k)}\|n'l'j')\mathbf{R}^{(SL)k}(nl, n'l') \quad \text{(IV-19a)}$$

where the sum is over nl, $n'l'$, j, j', S and L; and

$$\mathbf{M}^{(k)} = ie \sum \begin{Bmatrix} \frac{1}{2} & l & j \\ \frac{1}{2} & l' & j' \\ S & L & k \end{Bmatrix} [j, j', S, L]^{1/2} \left[\frac{(k+1)}{k(2k+1)} \right]^{1/2}$$

$$\times (nlj \| r^{-k-1}(\alpha \mathbf{C}^{(k)})^{(k)} \| n'l'j') R^{(SL)k}(nl, n'l') \quad \text{(IV-19b)}$$

where the sum is over nl, $n'l'$, j, j', S and L.

In this chapter, we are interested primarily in the treatment of the hyperfine interaction using first order perturbation theory. This means that we are concerned only with the parts of $\mathbf{Q}^{(k)}$ and $\mathbf{M}^{(k)}$ for which $nl = n'l'$; we shall designate these terms as $\mathbf{Q}_D^{(k)}$ and $\mathbf{M}_D^{(k)}$. The reduced matrices appearing in Eqs. (IV-19) take on a special symmetry when $nl = n'l'$ which enables us to truncate the sum over S and L when we speak of $\mathbf{Q}_D^{(k)}$ and $\mathbf{M}_D^{(k)}$. Consider the sum over j and j' in $\mathbf{Q}_D^{(k)}$. There will be one term in the sum with $j = j_1$, $j' = j_2$; another term in the sum will have $j = j_2$, $j' = j_1$. The reduced matrix in this second term in the sum will differ from the reduced matrix of the first term by only a phase factor, $(-1)^{j_1 - j_2}$. The $9-j$ symbols in the two terms will differ by a phase factor $(-1)^{j_1 + j_2 + 1 + S + L + k}$. Thus, since k is even for the $\mathbf{Q}^{(k)}$, the two terms in the sum will cancel unless $L + S$ is even. By applying this argument to all possible pairs, j, j', we see that $\mathbf{Q}_D^{(k)}$ vanishes identically unless $L + S$ is even. In the case of the magnetic multipoles $\mathbf{M}_D^{(k)}$, the pairs of terms will still differ by the phase $(-1)^{S+L+k}$, but because k is odd for magnetic multipoles, the pairs will cancel in this case unless $L + S$ is odd. We should emphasize, however, that these restrictions hold true only for $\mathbf{Q}_D^{(k)}$ and $\mathbf{M}_D^{(k)}$, and not for $\mathbf{Q}^{(k)}$ and $\mathbf{M}^{(k)}$.

The strength of the terms in this expansion decreases rapidly as k increases. The effects of only the first few terms in the series (IV-18) have actually been observed experimentally. We shall, in the future, confine our attention to the most important observable terms, which are the magnetic dipole and octupole interactions, $k = 1$ and 3 respectively, and the electric quadrapole interaction, $k = 2$. The explicit forms for these interactions are given below. For the magnetic dipole term we have[24]

$$\mathbf{M}_D^{(1)} = \sum_{nl} \left(\frac{2\mu_0}{3} [2(2l+1)]^{1/2} \left\langle \frac{\delta(r)}{r^2} \right\rangle_{10} R^{(10)1}(nl, nl) \right.$$

$$+ \frac{2\mu_0}{3} [6l(l+1)(2l+1)]^{1/2} \left\langle \frac{1}{r^3} \right\rangle_{01} R^{(01)1}(nl, nl)$$

$$\left. + 2\mu_0 \left[\frac{l(l+1)(2l+1)}{(2l+3)(2l-1)} \right]^{1/2} \left\langle \frac{1}{r^3} \right\rangle_{12} R^{(12)1}(nl, nl) \right) \quad \text{(IV-20)}$$

where

$$\left\langle \frac{\delta(r)}{r^2} \right\rangle_{10} = \frac{e}{\mu_0} \frac{1}{(2l+1)^2} [(l+1)^2 P_{++} + 2l(l+1)P_{+-} + l^2 P_{--}]$$

$$\left\langle \frac{1}{r^3} \right\rangle_{01} = -\frac{e}{\mu_0} \frac{1}{(2l+1)^2} [-(l+1)P_{++} + P_{+-} + lP_{--}]$$

$$\left\langle \frac{1}{r^3} \right\rangle_{12} = -\frac{e}{\mu_0} \frac{1}{3(2l+1)^2} [2(l+1)(2l-1)P_{++} - (2l-1)$$
$$\times (2l+3)P_{+-} + 2l(2l+3)P_{--}].$$

The P's are radial integrals defined by the equation

$$P_{jj'} = \int \frac{F_{nlj} G_{nlj'} + G_{nlj} F_{nlj'}}{r^2} \, dr;$$

in Eq. (IV-20) we have used symbols $\pm$ to represent $j = l \pm \frac{1}{2}$.

We next consider the quadrupole term $\mathbf{Q}_D^{(2)}$. It can be expressed as[24]

$$\mathbf{Q}_D^{(2)} = \sum_{nl} \left(e \left[\frac{l(l+1)(2l+1)}{6} \right]^{1/2} \left\langle \frac{1}{r^3} \right\rangle_{11} \mathbf{R}^{(11)2}(nl, nl) \right.$$
$$- e \left[\frac{l(l+1)(2l+1)(l+2)(l-1)}{56(2l+3)(2l-1)} \right]^{1/2} \left\langle \frac{1}{r^3} \right\rangle_{13} \mathbf{R}^{(13)2}(nl, nl)$$
$$\left. + e \left[\frac{(2l)(l+1)(2l+1)}{5(2l-1)(2l+3)} \right]^{1/2} \left\langle \frac{1}{r^3} \right\rangle_{02} \mathbf{R}^{(02)2}(nl, nl) \right) \quad \text{(IV-21)}$$

where

$$\left\langle \frac{1}{r^3} \right\rangle_{11} = -\frac{2\sqrt{6}}{5(2l+1)^2} [-(l+2)T_{++} + 3T_{+-} + (l-1)T_{--}]$$

$$\left\langle \frac{1}{r^3} \right\rangle_{13} = \frac{4\sqrt{21}}{5(2l+1)^2} [(2l-1)T_{++} + 4T_{+-} - (2l+3)T_{--}]$$

$$\left\langle \frac{1}{r^3} \right\rangle_{02} = \frac{1}{(2l+1)^2} [(2l-1)(l+2)T_{++} + 6T_{+-} + (l-1)(2l+3)T_{--}].$$

The $T_{jj'}$'s are radial integrals given by

$$T_{jj'} = \int \frac{F_{nlj} F_{nlj'} + G_{nlj} G_{nlj'}}{r^3} \, dr.$$

Finally, we have for the magnetic octupole term:

$$\mathbf{M}_D^{(3)} = -\mu_0 \sum_{nl} \left(\left[\frac{8l(l+1)(l-1)(2l+1)(l+2)}{7(2l+3)(2l-1)} \right]^{1/2} \left\langle \frac{1}{r^5} \right\rangle_{03} \mathbf{R}^{(03)3}(nl, nl) \right.$$

$$+ \left[\frac{32l(l+1)(2l+1)}{147(2l+3)(2l-1)} \right]^{1/2} \left\langle \frac{\delta(r)}{r^4} \right\rangle_{12} \mathbf{R}^{(12)3}(nl, nl)$$

$$\left. + \left[\frac{9(l-1)(l)(2l+2)(2l+1)(l+2)}{(2l+3)(2l+5)(2l-1)(2l-3)} \right]^{1/2} \left\langle \frac{1}{r^5} \right\rangle_{14} \mathbf{R}^{(14)3}(nl, nl) \right) \quad \text{(IV-22)}$$

where

$$\left\langle \frac{\delta(r)}{r^4} \right\rangle_{12} = \frac{e}{\mu_0} \frac{1}{2(2l+1)^2} [3(l+1)(l+2)U_{++} + 3l(l-1)U_{--} + 2(l-1)(l+2)U_{+-}]$$

$$\left\langle \frac{1}{r^5} \right\rangle_{03} = \frac{e}{\mu_0} \frac{1}{(2l+1)^2} [(l+1)U_{++} - lU_{--} - U_{+-}]$$

$$\left\langle \frac{1}{r^5} \right\rangle_{14} = -\frac{e}{\mu_0} \frac{1}{21(2l+1)^2} [2(2l-3)(2l+2)U_{++} + 4l(2l+5)U_{--} - (2l-3)(2l+5)U_{+-}].$$

The quantities $U_{jj'}$ appearing above are defined as

$$U_{jj'} = \int \frac{F_{nlj} G_{nlj'} + G_{nlj} F_{nlj'}}{r^4} dr.$$

Following the discussion of Section III-6, we see that Eqs. (IV-20)–(IV-22) can be viewed as defining either relativistic operators or effective operators. In the latter case, the substitution $\mathbf{R}^{(SL)K}(nl, nl) \to \mathbf{W}^{(SL)K}(nl, nl)$ must be made. The effective operators can then be evaluated between non-relativistic wavefunctions to give a relativistic result. In the former case, the operator is in the proper form for use when the wavefunction is composed of relativistic (sl) states.

4. EFFECTIVE OPERATORS

Let us consider the effective operators obtained by replacing $\mathbf{R}^{(SL)k}(nl, nl)$ by $\mathbf{W}^{(SL)k}(nl, nl)$ in the results of Section IV-3. It is often convenient to re-express these equivalent operators using the following equalities which are valid for matrix elements in the states of l^N:

$$\mathbf{W}^{(10)}(nl, nl) = \left(\frac{2}{[l]} \right)^{1/2} \mathbf{S},$$

$$W^{(01)}(nl, nl) = \left(\frac{3}{(2l)(l+1)(2l+1)}\right)^{1/2} \mathbf{L},$$

$$W^{(12)}(nl, nl) = -\left(\frac{10(2l-1)(2l+3)}{l(l+1)(2l+1)}\right)^{1/2} \sum_i (\mathbf{s}_i \mathbf{C}_i^{(2)}),$$

$$W^{(11)}(nl, nl) = \left(\frac{6}{l(l+1)(2l+1)}\right)^{1/2} \sum_i (\mathbf{s}_i \mathbf{l}_i),$$

$$W^{(02)}(nl, nl) = -\left(\frac{5(2l+3)(2l-1)}{(2l)(l+1)(2l+1)}\right)^{1/2} \sum_i \mathbf{C}_i^{(2)},$$

$$W^{(13)}(nl, nl) = -\left(\frac{56(2l+3)(2l-1)}{(l)(l+1)(2l+1)(l+2)(l-1)}\right)^{1/2} \sum_i (\mathbf{s}_i(\mathbf{C}_i^{(4)}\mathbf{l}_i)^{(3)}),$$

$$W^{(03)}(nl, nl) = -\left(\frac{14(2l+3)(2l-1)}{(l)(l+1)(2l+1)(l+2)(l-1)}\right)^{1/2} \sum_i (\mathbf{C}_i^{(4)}\mathbf{l}_i)^{(3)},$$

$$W^{(14)}(nl, nl) = \left(\frac{8(2l+5)(2l+3)(2l-1)(2l-3)}{(l+2)(l+1)(l)(l-1)(2l+1)}\right)^{1/2} \sum_i (\mathbf{C}_i^{(4)}\mathbf{s}_i).$$

The above results can easily be obtained using the Wigner-Eckart theorem.

Using these replacements, we find we can write the effective operators as

$$\mathbf{M}_D^{(1)} = \mu_0 \sum_i \left(\frac{4}{3}\left\langle\frac{\delta(r_i)}{r_i^2}\right\rangle_{10} \mathbf{s}_i + 2\left\langle\frac{1}{r_i^3}\right\rangle_{01} \mathbf{l}_i - 2\sqrt{10}\left\langle\frac{1}{r_i^3}\right\rangle_{12} (\mathbf{s}_i \mathbf{C}_i^{(2)})^{(1)}\right)$$

$$\mathbf{Q}_D^{(2)} = e \sum_i \left(\left\langle\frac{1}{r_i^3}\right\rangle_{11} (\mathbf{s}_i \mathbf{l}_i)^{(2)} + \left\langle\frac{1}{r_i^3}\right\rangle_{13} (\mathbf{s}_i(\mathbf{C}_i^{(4)}\mathbf{l}_i)^{(3)})^{(2)} - \left\langle\frac{1}{r_i^3}\right\rangle_{02} \mathbf{C}_i^{(2)}\right).$$

$$\mathbf{M}_D^{(3)} = \mu_0 \sum_i \left(4\left\langle\frac{1}{r_i^5}\right\rangle_{03} (\mathbf{C}_i^{(4)}\mathbf{l}_i)^{(3)} - 12\left\langle\frac{1}{r_i^5}\right\rangle_{14} (\mathbf{C}_i^{(4)}\mathbf{s}_i)^{(3)}\right.$$

$$\left. + \frac{8}{7}\sqrt{\left(\frac{5}{3}\right)}\left\langle\frac{\delta(r_i)}{r_i^4}\right\rangle_{12} (\mathbf{C}_i^{(2)}\mathbf{s}_i)^{(3)}\right). \quad \text{(IV-23)}$$

In this form, the operators bear the greatest resemblence to the classical interactions of Section I-2. Differences between these operators and those of Section I-2 will be discussed further in Section IV-6.

5. INTERACTION CONSTANTS AND NUCLEAR MOMENTS

The rotation group in three dimensions is, in many ways, a very well-behaved group. One of its more useful properties becomes evident when we study the Kronecker product of two representations $\mathscr{D}_L$ and $\mathscr{D}_{L'}$ of $R(3)$; the corresponding Clebsch-Gordan series is given by

$$\mathscr{D}_L \times \mathscr{D}_{L'} = \sum_{L''=|L-L'|}^{L+L'} \mathscr{D}_{L''}.$$

That is, $C_{L''}$ (see Section III-7) is always equal to either 0 or 1 for $R(3)$; it can never be greater than 1. This characteristic of $R(3)$ is of interest to us here because the Wigner-Eckart theorem assumes a simple form when $C_{L''} = 0$ or 1 only. In particular, the sum on the right-hand side of Eq. (III-43b) contains only one term, which implies that the tensor operators of rank k in electron space, $\mathbf{Q}^{(k)}$ and $\mathbf{M}^{(k)}$, can be replaced by other tensors of rank k acting in the space of the electrons. An equivalent statement can be made concerning the nuclear operators $\mathbf{F}^{(k)}$ and $\mathbf{N}^{(k)}$. Of course, the Wigner-Eckart theorem does not guarantee that the proportionality constant B_a will not be zero or infinity. To prevent having such an unpleasant proportionality constant, it is wise to replace a tensor operator with another tensor operator which shares as many properties as possible with the original operator. Thus, for instance, we would replace $\mathbf{M}_D^{(1)}$, a tensor operator of rank 1 having positive parity, with an angular momentum operator, which has the same properties, rather than with $\mathbf{r}$, which is a tensor of rank 1 with odd parity.

It was noted in Section IV-1 that neither I nor J is, strictly speaking, a good quantum number when hyperfine interactions are considered, but that it is an excellent approximation to assume that I is always a good quantum number. In addition, there are numerous cases when fine structure levels of different J are very widely separated, and to a very good approximation we can also consider J to be a good quantum number. In such cases, the Wigner-Eckart theorem can be used to express the hyperfine interactions in terms of operators composed of powers of $\mathbf{I}$ and $\mathbf{J}$. Thus, for the magnetic dipole interaction, matrix elements of $\mathbf{M}_D^{(1)}$ must be proportional to those of $\mathbf{J}$, the total electronic angular momentum; and matrix elements of $\mathbf{N}^{(1)}$ must be proportional to those of $\mathbf{I}$, the total nuclear angular momentum. We can then write

$$(\psi_N \psi_E | \mathbf{N}^{(1)} \cdot \mathbf{M}_D^{(1)} | \psi_N \psi_E) = A(\psi_N \psi_E | \mathbf{I} \cdot \mathbf{J} | \psi_N \psi_E).$$

The Wigner-Eckart theorem requires that A be independent of the M_I and M_J values of the states ψ_N and ψ_E. For convenience, we can evaluate the matrix elements above for the states $|IM_I = I)$ and $|JM_J = J)$, which we abbreviate as $|II)$ and $|JJ)$, respectively. In this case the matrix element on the right above is easily seen to be equal to IJ; that on the left is equal to $(II|N_0^{(1)}|II)(JJ|M_{D0}^{(1)}|JJ)$. Thus the proportionality constant A is given by

$$A = \frac{1}{IJ}(II|N_0^{(1)}|II)(JJ|M_{D0}^{(1)}|JJ)$$

$$= \left(\frac{\mu_I}{I}\right)[(J)(J+1)(2J+1)]^{-1/2}(J\|M_D^{(1)}\|J), \qquad \text{(IV-24)}$$

and the magnetic dipole hyperfine interaction can be written, within a manifold of states arising from a given I and J, as

$$H_D = A\mathbf{I} \cdot \mathbf{J}. \tag{IV-25}$$

In writing the second form of Eq. (IV-24), we have introduced the nuclear dipole moment, μ_I, which is defined by the equation

$$\mu_I = (II|N_0^{(1)}|II) = \beta_N(II|\sum_n (g_{ln}L_{nz} + g_{sn}S_{nz})|II). \tag{IV-26}$$

It can easily be seen that $\mathbf{N}^{(1)}$ is just the classical magnetic dipole moment of a current distribution provided by moving charged particles.[34] Because of definition (IV-26), $\mathbf{N}^{(1)}$ is related to μ_I through

$$\mathbf{N}^{(1)} = \frac{\mu_I}{I}\mathbf{I}.$$

We shall consider the physical interpretation of $(JJ|M_{D0}^{(1)}|JJ)$ in the next section.

Working under the same restrictions that I and J both be good quantum numbers, we can relate matrix elements of $\mathbf{Q}_D^{(2)}$ to those of $(\mathbf{JJ})^{(2)}$, and matrix elements of $\mathbf{F}^{(2)}$ to those of $(\mathbf{II})^{(2)}$. Thus

$$(\psi_N\psi_E|\mathbf{F}^{(2)} \cdot \mathbf{Q}_D^{(2)}|\psi_N\psi_E) = B(\psi_N\psi_E|(\mathbf{II})^{(2)} \cdot (\mathbf{JJ})^{(2)}|\psi_N\psi_E).$$

As before, B must be independent of the M_I and M_J values of the states, and we may determine B by evaluating these matrix elements using states having $M_I = I$, $M_J = J$. The matrix element on the right is easily evaluated using the recoupling equation

$$(\mathbf{IJ})^{(0)} \cdot (\mathbf{IJ})^{(0)} = \sum_k \begin{Bmatrix} 1 & 1 & 0 \\ 1 & 1 & 0 \\ k & k & 0 \end{Bmatrix} [k]^{1/2}(-1)^k(\mathbf{II})^{(k)} \cdot (\mathbf{JJ})^{(k)}.$$

Solving this equation for $(\mathbf{II})^{(2)} \cdot (\mathbf{JJ})^{(2)}$, we find

$$(\mathbf{II})^{(2)} \cdot (\mathbf{JJ})^{(2)} = (\mathbf{I} \cdot \mathbf{J})(\mathbf{I} \cdot \mathbf{J}) - \tfrac{1}{3}(\mathbf{I} \cdot \mathbf{I})(\mathbf{J} \cdot \mathbf{J}) + \tfrac{1}{2}(\mathbf{I} \cdot \mathbf{J}).$$

Utilizing the above result, one obtains in a straightforward manner

$$B = \frac{6(II|F_0^{(2)}|II)(JJ|Q_{D0}^{(2)}|JJ)}{I(2I-1)J(2J-1)}. \tag{IV-27a}$$

It is customary to define a second constant

$$b = \tfrac{2}{3}BIJ(2I-1)(2J-1)$$
$$= 4(II|F_0^{(2)}|II)(JJ|Q_{D0}^{(2)}|JJ) \tag{IV-27b}$$

such that quadrupole hyperfine interaction is described within a manifold of states arising from a J level by

$$H_Q = \frac{b[3(\mathbf{I}\cdot\mathbf{J})^2 + \tfrac{3}{2}(\mathbf{I}\cdot\mathbf{J}) - I(I+1)J(J+1)]}{2IJ(2I-1)(2J-1)}. \quad \text{(IV-28)}$$

We can further introduce the quantities

$$q_J = \frac{2}{e}(JJ|Q^{(2)}_{D0}|JJ) = \frac{2}{e}\left[\frac{(2J)(2J-1)}{(2J+3)(2J+2)(2J+1)}\right]^{1/2}(J\|Q^{(2)}_D\|J)$$

and

$$Q = \frac{2}{e}(II|F^{(2)}_0|II) = 2(II|\sum_n g_{ln} r_n^2 C^{(2)}_{n0}|II). \quad \text{(IV-29)}$$

The constant b is expressed in terms of these quantities by

$$b = e^2 q_J Q. \quad \text{(IV-30)}$$

Q is defined as the nuclear quadrupole moment. The expression $(2/e)\mathbf{F}^{(2)}$ is the classical quadrupole moment operator of the nucleus;[34,35] it can be related to Q through the relationship

$$\frac{2}{e}\mathbf{F}^{(2)} = \frac{\sqrt{(6)}Q}{I(2I-1)}(\mathbf{II})^{(2)}.$$

We shall investigate the significance of q_J in the following section.

By a similar but somewhat more tedious calculation, we find that the octupole interaction within a manifold of states arising from a given J can be expressed by the operator

$$H_O = c\{10(\mathbf{I}\cdot\mathbf{J})^3 + 20(\mathbf{I}\cdot\mathbf{J})^2 + 2(\mathbf{I}\cdot\mathbf{J})[-3I(I+1)J(J+1) + I(I+1)$$
$$+ J(J+1) + 3] - 5I(I+1)J(J+1)\}[I(I-1)(2I-1)(J)(J-1)(2J-1)]^{-1}$$

$$\text{(IV-31)}$$

where

$$c = (II|N^{(3)}_0|II)(JJ|M^{(3)}_{D0}|JJ)$$
$$= -\Omega\left[\frac{(J)(2J-1)(J-1)}{(J+1)(J+2)(2J+3)(2J+1)}\right]^{1/2}(J\|M^{(3)}_D\|J). \quad \text{(IV-32)}$$

Schwartz[23] defines the magnetic octupole moment of the nucleus, Ω, as

$$\Omega = -(II|N^3_0|II). \quad \text{(IV-33)}$$

The phase factor is introduced into Eq. (IV-33) so that a nucleus with positive dipole moment is likely to also have a positive octupole moment.

Equations (IV-25), (IV-28) and (IV-31) express the hyperfine interaction in a form which is very useful when the total electronic-nuclear wavefunction is constructed such that F is an eigenvalue, that is

$$|IJFM_F\rangle = \sum_{\substack{M_I \\ M_J}} [F]^{1/2}(-1)^{J-I-M_F} \begin{pmatrix} I & J & F \\ M_I & M_J & -M_F \end{pmatrix} |\alpha IM_I\rangle |\beta JM_J\rangle \quad \text{(IV-34)}$$

where α and β are any additional labels necessary to completely specify the nuclear and electronic states. Then we can write

$$(\mathbf{F} \cdot \mathbf{F}) = (\mathbf{I} + \mathbf{J}) \cdot (\mathbf{I} + \mathbf{J})$$

or

$$\mathbf{I} \cdot \mathbf{J} = \tfrac{1}{2}(\mathbf{F}^2 - \mathbf{J}^2 - \mathbf{I}^2)$$

where the operator on the right above has the eigenvalues

$$\tfrac{1}{2}(F(F+1) - J(J+1) - I(I+1))$$

when applied to a wavefunction such as given by Eq. (IV-33).

Use of the above result in Eq. (IV-25) shows that an "interval rule" exists for the dipole hyperfine interaction. That is, when only the dipole interaction is allowed by for example angular selection rules, the difference in energy between a state described by the quantum numbers (IJF) and a state described by $(IJF-1)$ is proportional to F:

$$E(IJF) - E(IJF-1) = FA.$$

This result is strictly valid, of course, only so long as J can be considered to be a good quantum number. However, in Chapter VII it is shown that if the breakdown of J is treated in second order perturbation theory, the result is simply to change the values of the interaction constants, but not the form of the interaction. Thus if the dipole hyperfine interaction is the only one allowed because of angular selection rules, the interval rule will not be affected by a small breakdown of J. (See, however, Section IX-2.) In addition, configuration interaction effects (Chapter V) cannot alter the F dependence of the energy separation.

Another interesting result also holds when I and J are both good quantum numbers. In this case, matrix elements of H_{hyp}, which we can expand as the dot product of tensor operators in the electronic $(\mathbf{T}_e^{(k)})$ and nuclear $(\mathbf{T}_n^{(k)})$ spaces, are all diagonal and all of the form

$$(IJFM|H_{\text{hyp}}|IJFM) = \sum_{k>0}(-1)^{I+J+F}\begin{Bmatrix} I & J & F \\ J & I & k \end{Bmatrix}(I\|T_n^{(k)}\|I)(J\|T_e^{(k)}\|J).$$

If these eigenvalues are multiplied by the degeneracy of the state $(2F+1)$ and a sum carried out over all possible values of F, the result (obtained using Eq.

(II-11)) is zero. That is, the center of gravity of a fine structure level is unchanged by the imposition of the hyperfine interaction. Deviations from this result can occur only because of a breakdown in J; however, for the reasons mentioned above, a breakdown in J small enough to be treated using second order perturbation theory will not cause any shifting of the center of gravity of the fine structure level.

6. NONRELATIVISTIC LIMITS OF THE HYPERFINE INTERACTIONS

The expressions discussed in the previous section relating to the first three terms in the expansion of the hyperfine interaction can be expanded in powers of $(v/c)^2$. The technique to be followed is identical to that utilized in the fine structure case; that is, F_{nlj} is replaced by R_{nl}, and G_{nlj} by

$$G_{nlj} = -\frac{\mu_0}{e}\left(\frac{d}{dr} - \frac{\kappa}{r}\right)R_{nl};$$

and $\mathbf{R}^{(Kk)}$ by $\mathbf{W}^{(Kk)}$ (see, however, the appendix).

For the magnetic dipole interaction, the radial integrals of interest are of the type

$$\int_0^\infty \frac{F_{nlj} G_{nlj'}}{r^2} dr = -\frac{\mu_0}{e}\int_0^\infty \frac{R_{nl}}{r^2}\left(\frac{d}{dr} - \frac{\kappa}{r}\right)R_{nl}\,dr$$

$$= (\kappa - 1)\frac{\mu_0}{e}\int_0^\infty \frac{R_{nl}^2}{r^3}\,dr - \frac{\mu_0}{e}\int_0^\infty \frac{R_{nl}}{r}\frac{d}{dr}\frac{R_{nl}}{r}\,dr$$

$$= +\frac{\mu_0}{2e}\left.\left|\frac{R_{nl}}{r}\right|^2\right|_{r=0} + (\kappa - 1)\frac{\mu_0}{e}\int_0^\infty \frac{R_{nl}^2}{r^3}\,dr.$$

Making the abbreviations

$$\int \frac{R_{nl}^2}{r^3}\,dr = \left\langle\frac{1}{r^3}\right\rangle \quad \text{and} \quad \left.\left|\frac{R_{nl}}{r}\right|^2\right|_{r=0} = \left\langle\frac{\delta(r)}{r^2}\right\rangle$$

we find that

$$P_{++} = \frac{2l\mu_0}{e}\left\langle\frac{1}{r^3}\right\rangle + \frac{\mu_0}{e}\left\langle\frac{\delta(r)}{r^2}\right\rangle$$

$$P_{--} = -2(l+1)\frac{\mu_0}{e}\left\langle\frac{1}{r^3}\right\rangle + \frac{\mu_0}{e}\left\langle\frac{\delta(r)}{r^2}\right\rangle$$

$$P_{+-} = P_{-+} = -\frac{\mu_0}{e}\left\langle\frac{1}{r^3}\right\rangle + \frac{\mu_0}{e}\left\langle\frac{\delta(r)}{r^2}\right\rangle.$$

Inserting these values into the radial parameters introduced by Eq. (IV-20), we find that, in the non-relativistic limit,

$$\left\langle \frac{1}{r^3} \right\rangle_{01} = \left\langle \frac{1}{r^3} \right\rangle, \quad \left\langle \frac{\delta(r)}{r^2} \right\rangle_{10} = \left\langle \frac{\delta(r)}{r^2} \right\rangle \quad \text{and} \quad \left\langle \frac{1}{r^3} \right\rangle_{12} = \left\langle \frac{1}{r^3} \right\rangle - \frac{1}{3}\left\langle \frac{\delta(r)}{r^2} \right\rangle.$$

Thus, in the non-relativistic limit, the magnetic dipole interaction Hamiltonian can be written as

$$H_D = \sum_{nl} \left(\frac{2}{3}\mu_0 [2(2l+1)]^{1/2} \left\langle \frac{\delta(r)}{r^2} \right\rangle \mathbf{W}^{(10)1} + \frac{2}{3}\mu_0 [6l(l+1)(2l+1)]^{1/2} \left\langle \frac{1}{r^3} \right\rangle \right.$$

$$\left. \times \mathbf{W}^{(01)1} + 2\mu_0 \left[\frac{l(l+1)(2l+1)}{(2l+3)(2l-1)} \right]^{1/2} \left[\left\langle \frac{1}{r^3} \right\rangle - \frac{1}{3}\left\langle \frac{\delta(r)}{r^2} \right\rangle \right] \mathbf{W}^{(12)1} \right) \cdot \mathbf{N}^{(1)}$$

(IV-35)

where we have abbreviated $\mathbf{W}^{(SL)}(nl, nl)$ as $\mathbf{W}^{(SL)}$. It is worthwhile at this point to consider the terms $\langle \delta(r)/r^2 \rangle$. For small values of r, the leading term in the non-relativistic radial wave-function[32] R_{nl} varies as r^{l+1}. Thus, $|R_{nl}/r^k|_{r=0}$ is zero for $k < l+1$, infinite for $k > l+1$, and has a finite value when $k = l+1$. Therefore, we see that $\langle \delta(r)/r^2 \rangle$ must be zero except when $l = 0$. However, matrix elements of $\mathbf{W}^{(12)1}$ must vanish for all states constructed with s electrons alone; this means that the $\langle \delta(r)/r^2 \rangle$ term multiplying the operator $\mathbf{W}^{(12)1}$ will never have any effect, and can be dropped. Once that is done, the non-relativistic Hamiltonian has the same form as the relativistic expression except that only two radial parameters are required in the non-relativistic limit while three are required in the relativistic case.

Using the identities of Section IV-4, we can write Eq. (IV-35) as

$$H_D = \frac{2\mu_I \mu_0}{I} \sum_i \left(\frac{2}{3}\left\langle \frac{\delta(r_i)}{r_i^2} \right\rangle \mathbf{s}_i + \left\langle \frac{1}{r_i^3} \right\rangle \mathbf{l}_i - \sqrt{10}\left\langle \frac{1}{r_i^3} \right\rangle (\mathbf{s}_i \mathbf{C}_i^{(2)})^{(1)} \right) \cdot \mathbf{I}. \quad \text{(IV-36)}$$

We have also used the results of the previous section to obtain the above expression. In order to get this result to agree with the results of Section I-2, one need only use the identity $\delta(r) = 4\pi\delta(\mathbf{r})r^2$. We can, at this point, easily understand the meaning of the non-relativistic form of the matrix element $(JJ|M_{D0}^{(1)}|JJ)$. Comparison of the value of $M_{D0}^{(1)}$ implicit in Eq. (IV-36) with the magnetic field operator of Section I-2 shows that $(JJ|M_{D0}^{(1)}|JJ)$ is, in the non-relativistic limit, simply the negative of the z component of the magnetic field at the origin produced by electrons in the state $|JJ\rangle$.

The non-relativistic limit of the interaction constant of Eq. (IV-24) is easily obtained using the above expressions:

$$A = \frac{2\mu_I \mu_0}{IJ} \langle JJ| \sum_i \left(\frac{8\pi}{3} \langle \delta(\mathbf{r}_i)\rangle s_{iz} + \left\langle \frac{1}{r_i^3}\right\rangle l_{iz} - \sqrt{10}\left\langle \frac{1}{r_i^3}\right\rangle (\mathbf{s}_i \, \mathbf{C}_i^{(2)})_0^{(1)}\right) |JJ\rangle$$

$$= \frac{2\mu_I \mu_0}{I} \langle J\| \sum_i \left(\frac{8\pi}{3} \langle \delta(\mathbf{r}_i)\rangle s_i + \left\langle \frac{1}{r_i^3}\right\rangle l_i - \sqrt{10}\left\langle \frac{1}{r_i^3}\right\rangle (\mathbf{s}_i \, \mathbf{C}_i^{(2)})^{(1)}\right) \|J\rangle$$

$$\times [J(J+1)(2J+1)]^{-1/2}. \qquad (IV-37)$$

In addition, it is sometimes convenient to write A in terms of three new constants

$$A = A_S + A_L + A_{sC} \qquad (IV-38)$$

where A_S is that portion of A proportional to the matrix element of $\mathbf{S}$; A_L, that proportional to the matrix element of $\mathbf{L}$; and A_{sC}, that proportional to the matrix element of $(\mathbf{sC}^{(2)})^{(1)}$. Obviously, these new constants can refer to either relativistic or non-relativistic dipole constants as Eq. (IV-23) or Eq. (IV-36) is used to describe the interaction.

It is interesting to compare Eq. (IV-36) and (IV-37) with the corresponding relativistically correct results obtained by using $M_D^{(1)}$ of Eq. (IV-23). Clearly the only difference between the two pairs of results is in the radial parameters. In the non-relativistic limit, both relativistic $\langle 1/r^3 \rangle$ parameters collapse to the same value. In addition, there is a very significant difference between $\langle \delta(r)/r^2 \rangle$ and $\langle \delta(r)/r^2 \rangle_{10}$. As explained above, $\langle \delta(r)/r^2 \rangle$ is zero except for s electrons. However, $\langle \delta(r)/r^2 \rangle_{10}$, although in general largest for s electrons, may be non-vanishing for electrons of any angular momentum. This implies that the non-relativistic A_S must be zero for all except s electrons, whereas the relativistic A_S can contribute for any electron.

The difference between $\langle \delta(r)/r^2 \rangle$ and $\langle \delta(r)/r^2 \rangle_{10}$ is generally of greatest importance when A_L and A_{sC} are required to vanish because of selection rules, and A_S need not. An example of such a case is given by the Hunds rule ground state of an atom having a half-filled shell of electrons. For example, in $Re(5d^5 \, ^6S_{5/2})$, both A_L and A_{sC} are zero because of selection rules in the orbital $R(3)$ space. The non-relativistic A_S must also be zero due to the vanishing of $\langle \delta(r)/r^2 \rangle$ for d electrons. The relativistic A_S need not vanish, however, and in fact has been calculated to be approximately $-122h$ Mc/sec.[67]

In the case of the electric quadrupole interaction, the radial integrals T_{++}, T_{+-}, and T_{--} all have the same limiting form to order $(v/c)^2$; all approach the value of $\langle 1/r^3 \rangle$. Using this limiting value, it can easily be shown that $\langle 1/r^3 \rangle_{11}$ and $\langle 1/r^3 \rangle_{13}$ both vanish and $\langle 1/r^3 \rangle_{02}$ becomes equal to $\langle 1/r^3 \rangle$. We then obtain

$$H_Q = e \sum_{nl} \left\langle \frac{1}{r^3}\right\rangle \left[\frac{(2l)(l+1)(2l+1)}{5(2l-1)(2l+3)}\right]^{1/2} \mathbf{W}^{(02)2} \cdot \mathbf{F}^{(2)} \qquad (IV-39a)$$

which can also be written as

$$H_Q = -e \sum_i \left\langle \frac{1}{r_i^3} \right\rangle \mathbf{C}_i^{(2)} \cdot \mathbf{F}^{(2)}. \tag{IV-39b}$$

In this limit, the parameter q_J defined in Eq. (IV-29) becomes

$$\begin{aligned} q_J &= -2 \langle JJ| \sum_i \left\langle \frac{1}{r_i^3} \right\rangle C_{i0}^{(2)} |JJ\rangle \\ &= -2 \langle J\| \sum_i \left\langle \frac{1}{r_i^3} \right\rangle C_i^{(2)} \|J\rangle \left[\frac{J(2J-1)}{(J+1)(2J+1)(2J+3)} \right]^{1/2}. \end{aligned} \tag{IV-40}$$

Comparison of this result with the classical results of Section I-2 shows that, in the non-relativistic limit, $-eq_J$ is the gradient in the z direction of the z component of the electric field at the origin produced by electrons in the state $|JJ\rangle$. The value of the constant b in the non-relativistic limit is obtained by multiplying q_J given above by $e^2 Q$.

The difference between the non-relativistic quadrupole interaction of Eqs. (IV-39) and the relativistic H_Q obtained from Eqs. (IV-21) and (IV-23) is more striking than the corresponding difference in the dipole case. Two of the angular terms in the relativistic quadrupole interaction vanish completely in the non-relativistic limit. This difference is generally of less importance than the differences which occur in the dipole case, however, because there are essentially no diagonal matrix elements for which $W^{(02)2}$ vanishes but $W^{(11)2}$ or $W^{(13)2}$ do not. As a result, there can be few cases analogous to the one discussed above concerning the dipole interaction, in which the entire interaction can be considered to arise from relativistic terms alone. Thus, because relativistic effects are generally small, expectation values of the relativistic and non-relativistic quadrupole interactions will most often be quite similar in value. The major exceptions to this statement can be expected to occur when all diagonal matrix elements of the quadrupole interaction disappear, for example in the half-filled shell (see Section IX-2), and off diagonal elements become dominant. We also note that the radial parameter which appears in the non-relativistic H_Q, $\langle 1/r^3 \rangle$, is the same one as appears in the non-relativistic dipole interaction. This, despite the fact that the relativistic radial parameters in the two cases are quite dissimilar.

Finally, we consider the magnetic octupole interaction. The radial integrals can again be reduced using an approach similar to that used for the magnetic dipole interaction. We find that, in the non-relativistic limit

$$U_{++} = (2l-2) \frac{\mu_0}{e} \left\langle \frac{1}{r^5} \right\rangle + \frac{\mu_0}{e} \left\langle \frac{\delta(r)}{r^4} \right\rangle$$

$$U_{--} = -(2l+4)\frac{\mu_0}{e}\left\langle\frac{1}{r^5}\right\rangle + \frac{\mu_0}{e}\left\langle\frac{\delta(r)}{r^4}\right\rangle$$

$$U_{+-} = -3\frac{\mu_0}{e}\left\langle\frac{1}{r^5}\right\rangle + \frac{\mu_0}{e}\left\langle\frac{\delta(r)}{r^4}\right\rangle$$

where

$$\int\frac{R_{nl}^2}{r^5}dr = \left\langle\frac{1}{r^5}\right\rangle \quad \text{and} \quad \left.\frac{R_{nl}^2}{r^2}\right|_{r=0} = \left\langle\frac{\delta(r)}{r^4}\right\rangle.$$

Using these values, one can easily show that

$$\left\langle\frac{1}{r^5}\right\rangle_{03} = \left\langle\frac{1}{r^5}\right\rangle, \quad \left\langle\frac{\delta(r)}{r^4}\right\rangle_{12} = \left\langle\frac{\delta(r)}{r^4}\right\rangle \quad \text{and} \quad \left\langle\frac{1}{r^5}\right\rangle_{14} = \left\langle\frac{1}{r^5}\right\rangle - \frac{1}{7}\left\langle\frac{\delta(r)}{r^4}\right\rangle.$$

Thus, in the non-relativistic limit, the octupole interaction can be written as

$$H_O = -\mu_0 \sum_{nl}\left(\left[\frac{8l(l+1)(l-1)(2l+1)(l+2)}{7(2l+3)(2l-1)}\right]^{1/2}\left\langle\frac{1}{r^5}\right\rangle W^{(03)3}\right.$$

$$+ \left[\frac{32l(l+1)(2l+1)}{3(49)(2l+3)(2l-1)}\right]^{1/2}\left\langle\frac{\delta(r)}{r^4}\right\rangle W^{(12)3}$$

$$+ \left.\left[\frac{9(l-1)(l)(2l+2)(2l+1)(l+2)}{(2l+3)(2l+5)(2l-1)(2l-3)}\right]^{1/2}\left\langle\frac{1}{r^5}\right\rangle W^{(14)3}\right)\cdot \mathbf{N}^{(3)}. \quad \text{(IV-41a)}$$

We have, in the above equation, dropped a term of the type $\langle\delta(r)/r^4\rangle W^{(14)3}$ because it will always have vanishing matrix elements for reasons analogous to those advanced for dropping a corresponding term in the dipole interaction. Utilizing the results of Section IV-4, we can also write H_O in the non-relativistic limit as

$$H_O = -\mu_0 \sum_i \left(12\left\langle\frac{1}{r_i^5}\right\rangle(\mathbf{C}_i^{(4)}\mathbf{s}_i)^{(3)} - 4\left\langle\frac{1}{r_i^5}\right\rangle(\mathbf{C}_i^{(4)}\mathbf{l}_i)^{(3)}\right.$$

$$\left. - \frac{8}{7}\sqrt{\frac{5}{3}}\left\langle\frac{\delta(r_i)}{r_i^4}\right\rangle(\mathbf{C}_i^{(2)}\mathbf{s}_i)^{(3)}\right)\cdot \mathbf{N}^{(3)}. \quad \text{(IV-41a)}$$

Making use of the techniques of tensor algebra (see e.g. Ref. 38, p. 150), and a result due to Innes and Ufford[68]

$$(\mathbf{C}^{(k+1)}\mathbf{l})^{(k)} = \left[\frac{k(2k-1)}{(k+1)(2k+3)}\right]^{1/2}(\mathbf{C}^{(k-1)}\mathbf{l})^{(k)},$$

one can show in a straightforward manner that, in the non-relativistic limit, $6(JJ|M_{D0}^{(1)}|JJ)$ is simply equal to the value of $\nabla_z^2 B_z$ at the origin produced by the electrons in the state $|JJ\rangle$. Here, B_z is the magnetic field produced by the electrons, given by $-M_{D0}^{(1)}$. In the same limit, the value of the constant c is given by

$$c = \Omega \mu_0 \langle JJ | \sum_i \left(12 \left\langle \frac{1}{r_i^5} \right\rangle (C_i^{(4)} s_i)_0^{(3)} - 4 \left\langle \frac{1}{r_i^5} \right\rangle (C_i^{(4)} l_i)_0^{(3)} \right.$$

$$\left. - \frac{8}{7} \sqrt{\frac{5}{3}} \left\langle \frac{\delta(r_i)}{r_i^4} \right\rangle (C_i^{(2)} s_i)_0^{(3)} \right) | JJ \rangle$$

$$= \Omega \mu_0 \langle J \| \sum_i \left(12 \left\langle \frac{1}{r_i^5} \right\rangle (C_i^{(4)} s_i)^{(3)} - 4 \left\langle \frac{1}{r_i^5} \right\rangle (C_i^{(4)} l_i)^{(3)} \right.$$

$$\left. - \frac{8}{7} \sqrt{\frac{5}{3}} \left\langle \frac{\delta(r_i)}{r_i^4} \right\rangle (C_i^{(2)} s)^{(3)} \right) \| J \rangle$$

$$\times \left[\frac{J(J-1)(2J-1)}{(J+1)(2J+1)(J+2)(2J+3)} \right]^{1/2} \quad \text{(IV-42)}$$

where Ω is defined by Eq. (IV-33).

Once again, the difference between relativistic and non-relativistic operators is primarily in the radial parameters. The two $\langle 1/r^5 \rangle$ values of the relativistic case are replaced by a single parameter in the non-relativistic equation. In addition, the non-relativistic $\langle \delta(r)/r^4 \rangle$ is non-vanishing for $p_{3/2}$ electrons only, while $\langle \delta(r)/r^4 \rangle_{12}$, although largest for $p_{3/2}$ electrons, may be non-zero for electrons of any angular momentum. At the present time, experimental evidence concerning the octupole interaction is so scarce that it is impossible to accurately judge the importance of relativistic effects, however.

V. The Central Field Approximation

1. INTRODUCTION

The hyperfine interaction of the previous chapter was studied using creation and annihilation operators for central-field wavefunctions. The results obtained in the latter part of the chapter are therefore based on the assumption that the states $|JM\rangle$ can be described in terms of sums of products of central-field wavefunctions for individual electrons. As discussed in Section III-10, a state $|JM\rangle$ constructed in this manner will be equivalent to the real atomic wavefunction only if an infinite number of electronic configurations are used to construct the state. In reality, of course, only a few configurations, usually one, are commonly utilized.

We shall discuss in this chapter the effects on hyperfine structure of such a truncation in the basis set which describes $|JM\rangle$. We shall, however, assume that all configurations that contribute strongly to the fine structure are included in $|JM\rangle$, so that our task will be only to study configurations which can be accurately treated with perturbation theory.

Before proceeding, we must consider in some detail the explicit form for the central potential to be used in the calculations of Chapter III. The choice of this potential is not a trivial matter, as the potential greatly influences the types of perturbations which must be considered in our discussion of hyperfine structure.

2. THE NONRELATIVISTIC CENTRAL POTENTIAL

We first consider the problem of a central potential for the non-relativistic calculations. A potential which is central and which can be expressed in a simple functional form has been given by Judd.[43] If we assume that our

The Central Field Approximation

interest lies in the configuration $\prod_k (n_k l_k)^{N_k}$, this potential is given by

$$U_c = \sum_{\alpha, \alpha'} |\alpha\rangle\langle\alpha| U_c |\alpha'\rangle\langle\alpha'| \qquad \text{(V-1a)}$$

where $|\alpha\rangle$ is a single particle state of the type $|n_\alpha l_\alpha m_{s\alpha} m_{l\alpha}\rangle$, and

$$\langle\alpha| U_c |\alpha'\rangle = \sum_\beta f(\beta)\left[\langle\alpha_1\beta_2| \frac{e^2}{r_{12}} |\alpha'_1\beta_2\rangle - \langle\alpha_1\beta_2| \frac{e^2}{r_{12}} |\beta_1\alpha'_2\rangle\right]. \qquad \text{(V-1b)}$$

The symbol β represents the single particle quantum numbers $(n_k l_k m_{sk} m_{lk})$. The sum over β runs over all possible values, and

$$f(\beta) = \frac{N_k}{4l_k + 2}.$$

The sum over β is effectively limited by the weighting factor $f(\beta)$ to those values of $n_k l_k$ which appear in the configuration of interest. It is important to note, however, that the sum over β extends over all possible m_s and m_l values in partially filled shells; these contributions are then "averaged" by the weighting factor $f(\beta)$.

A matrix element $\langle\alpha| U_c |\alpha'\rangle$ can be evaluated in a straightforward manner in order to demonstrate that U_c is indeed a central potential:

$$\langle\alpha| U_c |\alpha'\rangle = \sum_{\substack{\beta, q \\ K}} f(\beta)[(-1)^{l_\alpha + l_k - m_\alpha - m_{lk} + q} \begin{pmatrix} l_\alpha & K & l'_\alpha \\ -m_\alpha & q & m'_\alpha \end{pmatrix}$$

$$\times \begin{pmatrix} l_k & K & l_k \\ -m_{lk} & -q & m_{lk} \end{pmatrix} \langle l_\alpha \| C^{(K)} \| l'_\alpha\rangle \langle l_k \| C^{(K)} \| l_k\rangle$$

$$\times R^K(n_\alpha l_\alpha, n_k l_k; n'_\alpha l'_\alpha, n_k l_k) \delta(m_{s\alpha}, m'_{s\alpha})$$

$$- (-1)^{l_\alpha + l_k - m_\alpha - m_{lk} + q} \begin{pmatrix} l_\alpha & K & l_k \\ -m_\alpha & q & m_{lk} \end{pmatrix} \begin{pmatrix} l_k & K & l'_\alpha \\ -m_{lk} & -q & m'_\alpha \end{pmatrix}$$

$$\times \langle l_\alpha \| C^{(K)} \| l_k\rangle \langle l_k \| C^{(K)} \| l'_\alpha\rangle R^K(n_\alpha l_\alpha, n_k l_k; n_k l_k, n'_\alpha l'_\alpha)$$

$$\times \delta(m_{s\alpha}, m_{sk}) \delta(m_{sk}, m'_{s\alpha})].$$

R^K is a radial integral defined by

$$R^K(n_\alpha l_\alpha, n_\beta l_\beta; n_\gamma l_\gamma, n_\delta l_\delta)$$
$$= e^2 \iint R_{n_\alpha l_\alpha}(r_1) R_{n_\gamma l_\gamma}(r_1) R_{n_\beta l_\beta}(r_2) R_{n_\delta l_\delta}(r_2) \frac{r_<^K}{r_>^{K+1}} dr_1 \, dr_2.$$

The sums over m_{lk} and q in the second set of 3-j symbols above can be carried out using Eq. (II-4); the sums over m_{lk} and q in the first set can be carried out in the same way after first using an equation obtained from Eq. (II-6)

$$\sum_{j_3} [j_3] \begin{pmatrix} l_1 & l_2 & j_3 \\ n_1 & -n_2 & m_3 \end{pmatrix} \begin{pmatrix} j_1 & j_2 & j_3 \\ m_1 & m_2 & m_3 \end{pmatrix} \begin{pmatrix} j_1 & j_2 & j_3 \\ l_1 & l_2 & l_3 \end{pmatrix} (-1)^{l_1+l_2+l_3+n_1+n_2+n_3}$$

$$= \begin{pmatrix} j_1 & l_2 & l_3 \\ m_1 & n_2 & -n_3 \end{pmatrix} \begin{pmatrix} l_1 & j_2 & l_3 \\ -n_1 & m_2 & n_3 \end{pmatrix}.$$

Thus

$$\langle \alpha | U_c | \alpha' \rangle = \sum_{n_k l_k} f(\beta) [[l_k] R^0(n_\alpha l_\alpha, n_k l_k; n'_\alpha l_\alpha, n_k l_k)$$

$$- [l_\alpha]^{-1} \sum_K \langle l_\alpha \| C^{(K)} \| l_k \rangle^2 R^K(n_\alpha l_\alpha, n_k l_k; n_k l_k, n'_\alpha l_\alpha)]$$

$$\times \delta(m_{s\alpha}, m'_{s\alpha}) \delta(l_\alpha, l'_\alpha) \delta(m_{l\alpha}, m'_{l\alpha}).$$

This result shows that U_c is actually a central potential, which can connect only states which differ in no quantum number other than n, the principal quantum number.

We can now consider the effect of the perturbation

$$H_p = \sum_{i<j} \frac{e^2}{r_{ij}} - \sum_i U_c(r_i) \quad \text{(V-2)}$$

on a wavefunction ψ_0 which is an eigenfunction for electrons moving in the central field $\sum_i U_c(r_i)$. This perturbation is of interest to us because the first order hyperfine structure of Chapter IV must be corrected for second order perturbations of the type

$$\sum_{\psi_1} \frac{1}{\Delta E} \langle \psi_0 | H_{\text{hyp}} | \psi_1 \rangle \langle \psi_1 | H_p | \psi_0 \rangle \quad \text{(V-3)}$$

where ψ_1 belongs to a configuration different from that of ψ_0. Since H_{hyp} is composed of one-electron operators, the only values of ψ_1 that contribute to the above second order term are those that differ from ψ_0 by the quantum numbers of just one electron. Thus when considering $H_p \psi_0$ we need only investigate mechanisms which change ψ_0 by an excitation of a single electron. In the language of second quantization, we wish to study

$$H_p' | \psi_0 \rangle = \sum a_\alpha^\dagger b_\beta^\dagger \langle \alpha_1 \beta_2 | \frac{e^2}{r_{12}} | \gamma_1 \beta_2 \rangle b_\beta c_\gamma | \psi_0 \rangle$$

$$+ \sum a_\alpha^\dagger b_\beta^\dagger \langle \alpha_1 \beta_2 | \frac{e^2}{r_{12}} | \beta_1 \gamma_2 \rangle c_\gamma b_\beta | \psi_0 \rangle$$

$$- \sum a_\alpha^\dagger \langle \alpha | U_c | \gamma \rangle c_\gamma | \psi_0 \rangle \quad \text{(V-4)}$$

where the symbol c refers to electrons of the type nl; a, to electrons $n'l'$; and b, to electrons $n''l''$. Obviously, all three terms above correspond to an

82   The Central Field Approximation

excitation $nl \to n'l'$, with the two-body terms containing an interaction with the passive state $n''l''$. In addition to extending over all magnetic quantum numbers, the sum effectively extends over all values of nl and $n''l''$ appearing in ψ_0, and over all possible values of $n'l'$ not corresponding to filled shells in ψ_0. Let us consider for the moment only those terms in this summation for which the shell $n''l''$ is completely filled. By working $b_\beta^\dagger$ to the right of b_β and utilizing $b_\beta^\dagger |\psi_0\rangle = 0$ for a filled shell of $n''l''$ electrons, these terms in Eq. (V-4) can be written as

$$\sum_\beta \langle \alpha_1 \beta_2 | \frac{e^2}{r_{12}} | \gamma_1 \beta_2 \rangle a_\alpha^\dagger c_\gamma | \psi_0 \rangle$$

$$-\sum_\beta \langle \alpha_1 \beta_2 | \frac{e^2}{r_{12}} | \beta_1 \gamma_2 \rangle a_\alpha^\dagger c_\gamma | \psi_0 \rangle$$

$$-\sum_\phi f(\phi) \left[\langle \alpha_1 \phi_2 | \frac{e^2}{r_{12}} | \gamma_1 \phi_2 \rangle a_\alpha^\dagger c_\gamma | \psi_0 \rangle - \langle \alpha_1 \phi_2 | \frac{e^2}{r_{12}} | \phi_1 \gamma_2 \rangle a_\alpha^\dagger c_\gamma | \psi_0 \rangle \right]$$

(V-5)

where we have used Eq. (V-1) to evaluate the matrix element $\langle \alpha | U_c | \gamma \rangle$. It can easily be seen that this result is true even if $nl = n''l''$. Obviously, when $\phi = \beta$, $f(\phi) = f(\beta) = 1$ because the $n''l''$ shell is filled, and the above terms cancel completely.

This result is shown graphically by stating that the three graphs

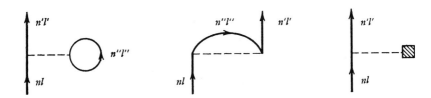

cancel, where the shaded square is that part of U_c arising from the $n''l''$ closed shell. The above graphs are drawn assuming that the nl shell is only partially filled. There is also cancellation between the graphs

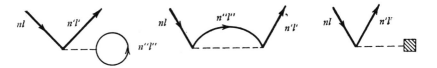

which apply when the shell nl is originally completely filled.

We have shown that in $H_p \psi_0$ there are no one-electron excitations due to interactions involving passive closed shells of the atom. There will be, however, one-electron excitations as a result of interactions involving open shells. To see that this is true, consider the case in which ψ_0 is given by a single determinental product state. We now assume that the shell $n''l''$ is only partially filled in ψ_0, and that $b_\beta^\dagger b_\beta |\psi_0\rangle$ is equal to $|\psi_0\rangle$ or 0 as the state $\beta = (n''l''m_{s\beta}m_{l\beta})$ is, or is not, present in $|\psi_0\rangle$. We then obtain an equation like Eq. (V-5), except that the sum over the magnetic quantum numbers $m_{s\beta}$ and $m_{l\beta}$ extends only over states which appear in ψ_0. The sum over $\phi = (n''l''m_{s\phi}m_{l\phi})$, on the other hand, is over all possible magnetic quantum numbers. There is consequently only an incomplete cancellation in this case.

A potential often used in atomic calculations is the Hartree-Fock potential, U_{HF}. A one-electron matrix element of U_{HF} is defined by[32]

$$\langle \alpha | U_{HF} | \alpha' \rangle = \sum_{\beta=1}^{N} \left[\langle \alpha_1 \beta_2 | \frac{e^2}{r_{12}} | \alpha'_1 \beta_2 \rangle - \langle \alpha_1 \beta_2 | \frac{e^2}{r_{12}} | \beta_1 \alpha'_2 \rangle \right]. \quad \text{(V-6)}$$

The eigenfunction in a Hartree-Fock calculation, ψ_{HF}, is always taken to be a single determinental product state. The sum over β in Eq. (V-6) runs over all one-electron states contained in this determinental product state. In the case in which all electrons are in closed shells, $U_c = U_{HF}$; in such a case, U_{HF} is also a central potential. However, when there are electrons in partially filled shells, U_{HF} is no longer central. This can easily be seen by considering the proof given above demonstrating that U_c is a central potential. This proof hinged on the statement that β ran over all possible values of $m_{l\beta}$ and $m_{s\beta}$, even in shells only partially filled. As a result, sums over magnetic quantum numbers could be carried out. For the Hartree-Fock potential, these sums generally cannot be carried out in partially filled shells, since the sums run over only those magnetic quantum numbers which appear in the state ψ_{HF}. One exception occurs when the partially filled shells in ψ_{HF} are half-filled, with all spins in each of the half-filled shells aligned. In this case, all $m_{l\beta}$ values are present, and the sums over the orbital magnetic quantum numbers can be carried out. However, in this case, matrix elements of U_{HF} will vary in value depending on the spin projection of the state used to evaluate the matrix element; this, of course, implies that U_{HF} is not truly central.

We can consider the effect of the operator

$$H_{pHF} = \sum_{i<j} \frac{e^2}{r_{ij}} - \sum_i U_{HF}(r_i)$$

on ψ_{HF}. The arguments proceed exactly as those above concerning $H_p\psi_0$. The result, however, is somewhat different. We find that, if H'_{pHF} is the operator analogous to H'_p of Eq. (V-4), that is, the operator describing one-electron excitations from the ground-state wavefunction ψ_{HF}, then

$$H'_{pHF} \psi_{HF} = 0.$$

In other words, the perturbation H_{pHF} produces no one-electron excitations from ψ_{HF}. Thus, if Hartree-Fock wavefunctions are used, a one-electron operator such as H_{hfs} will have no second order energy shifts of the type (V-3); the first non-vanishing perturbation would be in third order.

Unfortunately, real Hartree-Fock wavefunctions are virtually impossible to obtain unless all electrons are in filled shells. If there are partially filled shells, the potential is no longer central, and the eigenfunctions can no longer be separated into radial and angular parts according to the usual separation

$$\psi = Y_{lm}(\theta, \phi) \frac{R_{nl}(r)}{r} \chi_{ms}^{\frac{1}{2}}. \tag{V-7}$$

Without such separation, solutions are exceedingly difficult to determine. Open shell "Hartree-Fock" calculations are in reality performed by, for example, forcing the eigenfunctions to be of the form given by Eq. (V-7). These wavefunctions, which are not really solutions of the Hartree-Fock equations, are often called restricted Hartree-Fock wavefunctions. When one uses wavefunctions of this type, there may indeed by one-electron excitations produced by action of H_{pHF} on ψ_{HF}. These excitations will primarily come from the interaction with passive particles in the partially filled shells, just as in the case discussed earlier concerning the excitations of wavefunctions calculated using U_c. Thus, we can expect perturbation results obtained using H_p to be similar qualitatively to those obtained using a H_{pHF} based on a restricted Hartree-Fock potential.

3. THE RELATIVISTIC CENTRAL FIELD

We can define a relativistic central field suitable for use in Eq. (III-4) by analogy with the results of the preceding section. If our interest is in the relativistic configuration $\prod_k (n_k l_k j_k)^{N_k}$, a relativistic central potential is given by

$$U_{cr} = \sum_{\alpha, \alpha'} |\alpha)(\alpha| U_{cr} |\alpha')(\alpha'| \tag{V-8a}$$

where α is a single-particle state of the type $|n_\alpha l_\alpha j_\alpha m_\alpha)$, and

$$(\alpha| U_{cr} |\alpha') = \sum_\beta g(\beta) \left[(\alpha_1 \beta_2 | \frac{e^2}{r_{12}} | \alpha'_1 \beta_2) - (\alpha_1 \beta_2 | \frac{e^2}{r_{12}} | \beta_1 \alpha'_2) \right]. \tag{V-8b}$$

Here, β is a state of the type $|n_k l_k j_k m_k)$, and

$$g(\beta) = \frac{N_k}{2j_k + 1}.$$

U_{cr} can easily be shown to be a central potential (i.e., in this case diagonal with respect to l, j and m) using the techniques of the previous section.

In addition, a relativistic analog of H_p can be formed

$$H_{pr} = \sum_{i<j} \frac{e^2}{r_{ij}} - \sum_i U_{cr}(r_i).$$

Evaluation of $H_{pr}|\psi_0\rangle$ where $|\psi_0\rangle$ is a relativistic wavefunction composed of electrons of the type (njl) can be carried out as in the corresponding non-relativistic case. One can easily show that H_{pr} can produce no one-electron excitations from $|\psi_0\rangle$ due to interactions involving passive electrons in closed shells. As in the non-relativistic case, however, there will be one electron excitations due to interaction in which the passive electron is in a partially filled shell.

If our interest lies in the relativistic configuration $\prod_k (n_k l_k)^{M_k}$, the central potential given above is not the most convenient one to consider. In this case, a convenient central potential suitable for use in Eq. (III-4) is defined by the matrix element

$$(\alpha|U_{cr}|\alpha') = \sum_\beta h(\beta)\left[(\alpha_1\beta_2|\frac{e^2}{r_{12}}|\alpha'_1\beta_2) - (\alpha_1\beta_2|\frac{e^2}{r_{12}}|\beta_1\alpha'_2)\right], \quad \text{(V-9)}$$

where $h(\beta) = (M_k/4l_k + 2)$, and the states α and β are as defined above. This potential can also easily be seen to be central. However, when considering H_{pr} defined using this potential, we find that not all one-electron excitations from $|\psi_0\rangle$ caused by interactions involving passive electrons in closed shells vanish. Only those excitations vanish for which not only the shell of the passive electron, but also the other j shell arising from the same l value, is filled.

In order to use U_{cr} in the (nl) scheme, e^2/r_{12} must be transformed according to Eqs. (III-34) and (III-35); we will refer to this transformed operator as $(e^2/r_{12})_{\text{trans}}$ for simplicity. In addition, the states appearing in Eq. (V-9) must be transformed using Eq. (III-17). The resulting form of U_{cr}, valid when dealing with configurations $\prod_k (n_k l_k)^{M_k}$ of relativistic electrons, is

$$U_{cr} = \sum_{\alpha,\alpha'} |\alpha)(\alpha|U_{cr}|\alpha')(\alpha'| \quad \text{(V-10)}$$

where, now $\alpha = |n_\alpha l_\alpha m_{s\alpha} m_{l\alpha})$ and

$$(\alpha|U_{cr}|\alpha') = \sum_\beta h(\beta)\left[(\alpha_1\beta_2|\left(\frac{e^2}{r_{12}}\right)_{\text{trans}}|\alpha'_1\beta_2) - (\alpha_1\beta_2|\left(\frac{e^2}{r_{12}}\right)_{\text{trans}}|\beta_1\alpha'_2)\right].$$

The states β are now relativistic states of the type $|n_k l_k m_{sk} m_{lk})$.

The transformed e^2/r_{ij} is not completely scalar in either the spin or the orbital space, but rather is scalar only in the total J space. One should not, therefore, expect U_{cr} to be diagonal in the quantum numbers m_s, m_l; that is, one should not expect U_{cr} to be a central potential for relativistic (nl) electrons. In fact, it is quite simple to show that U_{cr} is not diagonal in these quantum numbers by performing a calculation similar to that done in the preceding section. This result should not disturb one unduly, however, since U_{cr} is a central potential in the (nlj) scheme, which is the one in which the wavefunctions must originally be calculated.

We can now consider H_{pr} in the (nl) scheme. Again, a calculation such as carried out in the previous section shows that there are no one-electron excitations caused by H_{pr} in which the passive electron is in closed shells. Interactions in which the passive electron is in an open shell are somewhat complicated, however, because many different angular dependencies arise in $(e^2/r_{ij})_{trans}$, as shown in Eq. (III-34). It can be demonstrated,[61] however, that only the terms behaving like $\sum_{i<j} e^2/r_{ij}$ and $\sum_{i<j} \mathbf{w}_i^{(11)0} \cdot \mathbf{w}_j^{(00)0}$ (the spin-orbit interaction) are of appreciable size. We shall henceforth consider only these larger terms when considering H_{pr} in the transformed scheme.

4. BREAKDOWN OF THE CENTRAL FIELD APPROXIMATION

As we saw in Chapter III, when using the central field approximation we were able to separate the electronic wavefunction into a radial part depending on n, l, and j, and an angular part depending on l, j and m. A non-relativistic approach would allow the wavefunction to be separated into a radial part depending on n and l, and an angular part depending on s, m_s, l and m_l. For the moment, we will discuss only the non-relativistic case as it is much easier to visualize results in this limit. Although there are many errors associated with these central field approximations, two in particular return to haunt us at this point. The first concerns the nature of the Coulomb exchange interaction. The Coulomb repulsion term between electrons, $e^2/r_{12} = e^2 \sum_k \mathbf{C}_1^{(k)} \cdot \mathbf{C}_2^{(k)} r_<^k/r_>^{k+1}$, can be evaluated between antisymmetrized states $\{\psi_1 \psi_2\}$ or, alternatively a modified Coulomb repulsion operator[19]

$$\sum_k \left\{ R^k(l_1, l_2; l_1, l_2) \mathbf{C}_1^{(k)} \cdot \mathbf{C}_2^{(k)} - 4\delta(m_{s1}, m_{s2}) \langle l_1 \| C^{(k)} \| l_2 \rangle^2 \right.$$
$$\left. \times R^k(l_1, l_2; l_2, l_1) \sum_K (-1)^K \begin{Bmatrix} l_1 & l_1 & K \\ l_2 & l_2 & k \end{Bmatrix} \mathbf{w}_1^{(0K)}(l_1, l_1) \cdot \mathbf{w}_2^{(0K)}(l_2, l_2) \right\}$$

can be evaluated between unsymmetrized states. Because of the $\delta(m_{s1}, m_{s2})$ term in this modified operator, an electron in an unfilled shell with, say, spin up, will interact differently with the electrons in the core with spin up than with the core electrons having spin down. This interaction tends to cause the core electrons with a given nl and spin up to be drawn closer to the spin-up electron in the unfilled shell (due to the negative sign of the $\delta(m_{s1}, m_{s2})$ term in the Coulomb interaction) than are core electrons of the same nl and spin down. Thus R_{nl} really becomes R_{nlm_s} for the core electrons. The second error of interest here strikes even more deeply at the concept of a central potential. An unfilled shell of electrons with $l > 0$ is not spherical; this non-spherical shell destroys the spherical symmetry of the potential in which all other electrons move, violating the basic requirements of the central field model. Without the central field, eigenfunctions of the Hamiltonian cannot be separated into angular and radial parts, and l is no longer a good quantum number for the electron. Both of these effects are closely related and can be treated equivalently using perturbation theory. The first is the main component of what is often called core polarization, which affects the dipole interaction constant, A. The second effect gives rise primarily to what is known as quadrupole, or Sternheimer, shielding, which affects the value of the quadrupole constant b.

Perturbation theory is, in general, a difficult technique to use in evaluating these effects because the sums over excited states generally do not converge quickly. However, the perturbation calculation does reveal a great deal about the form of the interaction and for that reason can be quite useful. We shall consider here the first-order change in matrix elements of the relativistic operator $\mathbf{T}^{(SL)J} = \sum_l D(l, l) \mathbf{R}^{(SL)J}(nl, nl)$ caused by a mixing of configurations by the Coulomb interaction. Higher-order terms are discussed in Section V-8. We shall restrict our attention to values of S, L, and J which appear in the hyperfine Hamiltonian.

This first-order change in matrix elements of the operator $\mathbf{T}^{(SL)J}$ due to coulomb-induced configuration interaction is written as

$$\sum_{\psi_i} \frac{(\psi_0 | T^{(SL)J}_{M_J} | \psi_i)(\psi_i | H_{pr} | \psi_0')}{E_0' - E_i} + \frac{(\psi_0 | H_{pr} | \psi_i)(\psi_i | T^{(SL)J}_{M_J} | \psi_0')}{E_0 - E_i}. \quad \text{(V-11)}$$

Because $\mathbf{T}^{(SL)J}$ is a one-electron operator, ψ_i can differ from ψ_0 by the quantum numbers of only one electron. We have already made two assumptions concerning the form of ψ_i and ψ_0 which bear re-emphasizing now. The first is that they are both eigenfunctions of the Hamiltonian H_{fs} (Eq. (III-44)). The second is that ψ_i is mixed only weakly with ψ_0 by H_{pr}. The type of perturbations which we shall consider henceforth also satisfies a final assumption: the separations between the states ψ_0 in the ground configuration C_0 and the separations between the states ψ_i in the excited configuration C_i are both much smaller than the distance between the configurations C_0 and C_i. Then, $E_0' - E_i \approx E_0 - E_i$ for all ψ_i, and $E_0 - E_i \approx \Delta E$ for all ψ_0, ψ_i. The

The Central Field Approximation

energy denominators can now be taken outside of the sum in the perturbation expression.

Having made these assumptions, we can obtain an operator acting in the space of C_0 which reproduces the second-order effects. This is done simply using the second quantization techniques of Chapter II.[43] First, the operators $\mathbf{T}^{(SL)J}$ and H_{pr} are expressed in second quantized form. Then, from the resulting sum over creation and annihilation operators we pick only those which can connect C_0 to C_i. Having done that, we can extend the sum $\sum_{\psi_i} |\psi_i\rangle\langle\psi_i|$, which is originally taken over all states of C_i, to a sum over all possible atomic states. This extension of the limits of the summation will not change the value of (V-11) because the creation and annihilation operators can connect only states of C_i to those of C_0. However, once ψ_i has been extended in this manner, we can invoke closure in order to replace $\sum_{\psi_i} |\psi_i\rangle\langle\psi_i|$ with 1. The annihilation and creation operators for $\mathbf{T}^{(SL)J}$ and H_{pr} can then be brought together, and the anticommutation relations (II-28), (II-30) and (II-31) can be used to simplify the resulting expression.

Let us discuss a simple example of such a calculation. Consider the perturbation

$$\frac{1}{\Delta E} \sum_{\psi_i} \langle\psi_0| T^{(SL)}_{M_S M_L} |\psi_i\rangle\langle\psi_i| \sum_{\substack{K \\ k<j}} 2[K]^{-1} C^K(ll_2; ll_1) \mathbf{r}_j^{(0K)}(l,l) \cdot \mathbf{r}_k^{(0K)}(l_2, l_1) |\psi_0\rangle \tag{V-12}$$

where we assume that

$$C_0 = l^N l_1^{4l_1+2}, \quad C_i = l^N l_1^{4l_1+1} l_2.$$

The term $2[K]^{-1} C^K(ll_2; ll_1)$ is the factor multiplying that part of $(e^2/r_{12})_{\text{trans}}$ having the angular dependence $\sum_{j<k} \mathbf{r}_j^{(0K)}(l,l) \cdot \mathbf{r}_k^{(0K)}(l_2, l_1)$. This term is simply $\sum_{j<k} e^2/r_{jk}$ expressed in relativistic form. Writing the above operators in second quantized notation and using the closure theorem, we obtain

$$\frac{1}{\Delta E}\langle\psi_0| \left[\sum \langle\alpha| t^{(SL)}_{M_S M_L} |\beta\rangle (\delta_1 \gamma_2 | \mathbf{r}_1^{(0K)}(l,l) \cdot \mathbf{r}_2^{(0K)}(l_2, l_1) |\varepsilon_1 \phi_2) \right.$$
$$\left. \times [K]^{-1} C^K(ll_2; ll_1) c_\alpha^\dagger d_\beta d_\gamma^\dagger b_\delta^\dagger b_\varepsilon c_\phi \right] |\psi_0\rangle$$

where the operators c refer to relativistic electrons of angular momentum l_1, d to electrons of angular momentum l_2, and b, to electrons of angular momentum l. The subscripted indices contain the remaining quantum numbers necessary to specify a single-particle state, and the sum is over all values of these indices and K. We have also used $T^{(SL)}_{M_S M_L} = \sum_i t^{(SL)}_{i M_S M_L}$. Because of the form of C_0, $c^\dagger|\psi_0) = d|\psi_0) = 0$; using the anticommutation relations, we therefore obtain

$$\frac{1}{\Delta E}(\psi_0|\left[\sum (\alpha|t^{(SL)}_{M_S M_L}|\beta)(\delta_1\gamma_2|\mathbf{r}^{(0K)}_1(l,1)\cdot\mathbf{r}^{(0K)}_2(l_2,l_1)|\varepsilon_1\phi_2)\right.$$

$$\times [K]^{-1}C^K(ll_2;ll_1)\,\delta(\alpha,\phi)\,\delta(\gamma,\beta)b_\delta^\dagger b_\varepsilon\bigg]|\psi_0)$$

$$=\frac{1}{2\Delta E}(\psi_0|\left[\sum\begin{pmatrix}l_1 & L & l_2\\-m_{l\alpha} & M_L & m_{l\beta}\end{pmatrix}\begin{pmatrix}l_2 & K & l_1\\-m_{l\beta} & q & m_{l\alpha}\end{pmatrix}\begin{pmatrix}l & K & l\\-m_{l\delta} & -q & m_{l\varepsilon}\end{pmatrix}\right.$$

$$\times\begin{pmatrix}\tfrac{1}{2} & S & \tfrac{1}{2}\\-m_{s\alpha} & M_s & m_{s\beta}\end{pmatrix}(\tfrac{1}{2}l_1\|t^{(SL)}\|\tfrac{1}{2}l_2)C^K(ll_2;ll_1)$$

$$\times(-1)^{l_1-m_{l\alpha}+l_2-m_{l\beta}+l-m_{l\delta}+\tfrac{1}{2}-m_{s\alpha}+q}\,\delta(m_{s\alpha},m_{s\beta})\,\delta(m_{s\delta},m_{s\varepsilon})b_\delta^\dagger b_\varepsilon\bigg]|\psi_0).$$

Carrying out some of the indicated summations, we obtain

$$(\psi_0|\sum_{\delta,\varepsilon}b_\delta^\dagger\left[\frac{1}{\Delta E}\begin{pmatrix}\tfrac{1}{2} & 0 & \tfrac{1}{2}\\-m_{s\delta} & 0 & m_{s\varepsilon}\end{pmatrix}\begin{pmatrix}l & L & l\\-m_{l\delta} & M_L & m_{l\varepsilon}\end{pmatrix}\right.$$

$$\times\,\delta(S,0)[L]^{-1}(-1)^{\tfrac{1}{2}-m_{s\delta}+l-m_{l\delta}}(\tfrac{1}{2}l_1\|t^{(SL)}\|\tfrac{1}{2}l_2)C^L(ll_2;ll_1)\bigg]b_\varepsilon|\psi_0).$$

The term in the outer square brackets is the matrix element of an operator P which acts only within the ground configuration C_0, where

$$P = \frac{1}{\Delta E}\frac{\delta(S,0)}{[L]^{\tfrac{1}{2}}}(\tfrac{1}{2}l_1\|t^{(SL)}\|\tfrac{1}{2}l_2)C^L(ll_2;ll_1)r^{(SL)}_{M_S M_L}(l,l).$$

The matrix element above is thus simply the matrix element of the second quantized form of P. We have therefore found an operator which acts in the ground configuration to reproduce the effect of the perturbation (V-12).

We shall carry out such calculations for all the terms in (V-11). First, however, it is useful to divide both H_{pr} and the states ψ_i into two sections each. The operator $(e^2/r_{12})_{\text{trans}}$ has two main components, as explained above. The first is e^2/r_{12} in the (nl) scheme; the second is the part of the spin-orbit interaction which arises from the interaction of the electron i with the gradient of the central potential produced by the other electrons in the atom. (The remainder of the spin-orbit interaction arises from the transformed form of $-Ze^2/r_i$). It will be convenient to consider these two parts separately. It is also convenient to split the sum over ψ_i into two portions; in the first C_i differs from C_0 by the raising of an electron from a closed shell to a shell other than one of the partially filled shells; in the second, C_i differs from C_0 by the raising of an electron from or to a partially filled shell.

a. Excitation of an Electron from a Filled to an Empty Shell

Consider first the former case for the intermediate configuration C_i and the part of H_{pr} arising from the effective e^2/r_{ij} only. We find, using the techniques illustrated above, that the operator $\mathbf{T}^{(SL)J}$ can be corrected for perturbations of this type simply by replacing the term $D(l, l)$ by $(1 + \Delta_1)D(l, l)$, where Δ_1 is given by Eq. (V-13) (Table V-1). In obtaining this equation,

Table V-1. Equations Necessary to Describe Perturbations Involving the Operator $\mathbf{T}^{(SL)J}$

$$\Delta_1 = \sum \frac{2D(l', l_1)}{\Delta E D(l, l)} \left[\frac{2\delta(S, 0)}{[L]} C^L(ll'; ll_1) + \sum_k (-1)^{L+k+1} \begin{Bmatrix} l & L & l \\ l_1 & k & l' \end{Bmatrix} C^k(l'l; ll_1) \right]. \quad \text{(V-13)}$$

$$\mathbf{N}^{(\kappa k)J} = \sum \begin{Bmatrix} S & 1 & \kappa \\ \tfrac{1}{2} & \tfrac{1}{2} & \tfrac{1}{2} \end{Bmatrix} \begin{Bmatrix} L & S & J \\ \kappa & k & 1 \end{Bmatrix} \left(\frac{[\kappa, k, 1, S, L]}{[\tfrac{1}{2}, l]} \right)^{1/2} \frac{D(l_1, l')}{\Delta E}$$

$$\times (-1)^{J+1}((-1)^{k+L} + (-1)^{S+\kappa})(A(l'l, ll_1)$$

$$+ A(ll'; l_1 l)) \begin{Bmatrix} L & 1 & k \\ l & l & l \end{Bmatrix} \mathbf{R}^{(\kappa k)J}(nl, nl). \quad \text{(V-15)}$$

$$\mathbf{M}^{(SL)J} = \sqrt{2} \sum \frac{1}{\Delta E} \left(\frac{[\kappa]}{[k]} \right)^{1/2} \Bigg\{ D(l', l_\phi) C^k(l_\alpha l_\beta ; l'l_\gamma)$$

$$\times \begin{Bmatrix} L & k & \kappa \\ l_\alpha & l_\phi & l' \end{Bmatrix} (-1)^{k+\kappa+l_\phi+l_\alpha} + D(l_\alpha, l') C^k(l_\beta l'; l_\gamma l_\phi)$$

$$\times \begin{Bmatrix} L & k & \kappa \\ l_\phi & l_\alpha & l' \end{Bmatrix} (-1)^{l_\alpha+l_\phi+L} \Bigg\} \sum_{i\ne j} (\mathbf{r}_i^{(Sk)}(n_\alpha l_\alpha, n_\phi l_\phi) \mathbf{r}_j^{(0k)}(n_\beta l_\beta, n_\gamma l_\gamma))^{(SL)J}. \quad \text{(V-19)}$$

$$\Delta_2 = \sum \frac{2h(l_1)}{\Delta E} \frac{D(l, l')}{D(l, l)} \left\{ C^0(l'l_1; ll_1) + \delta(l, l')(-1)^{l+l_1+1} \sum_k [l]^{-1} C^k(l'l_1 : l_1 l) \right\}. \quad \text{(V-20)}$$

$$\mathbf{P}^{(\kappa k)J} = \sum \frac{1}{\Delta E} \begin{Bmatrix} S & 1 & \kappa \\ \tfrac{1}{2} & \tfrac{1}{2} & \tfrac{1}{2} \end{Bmatrix} \begin{Bmatrix} \kappa & k & J \\ L & S & 1 \end{Bmatrix} \begin{Bmatrix} 1 & L & k \\ l & l_\alpha & l_\alpha \end{Bmatrix} A(l_1 l'; l_\beta l_\alpha)$$

$$\times D(l, l')[k, \kappa, S, L, 1]^{1/2}(-1)^{l_\alpha+l+J}$$

$$\times [(-1)^{S+\kappa} + (-1)^{L+k}] \sum_{i\ne j} \mathbf{r}_i^{(00)0}(n_1 l_1, n_\beta l_\beta) \mathbf{r}_j^{(\kappa k)J}(nl, n_\alpha l_\alpha). \quad \text{(V-21a)}$$

$$\mathbf{T}^{(\kappa k)J} = \sum \frac{2}{\Delta E} \frac{D(l, l')}{[\tfrac{1}{2}, l_\alpha]^{1/2}} A(l'l_1; l_\alpha l_\beta) \begin{Bmatrix} \kappa & k & J \\ L & S & 1 \end{Bmatrix} \left(\frac{[\kappa, k]}{[1]} \right)^{1/2} (-1)^{k+S+J+1}$$

$$\times \sum_{i\ne j} (\mathbf{r}_i^{(SL)}(nl, n_\alpha l_\alpha) \mathbf{r}_j^{(11)}(n_1 l_1, n_\beta l_\beta))^{(\kappa k)J}. \quad \text{(V-21b)}$$

Table V-1 (*Continued*)

$$\mathbf{P}^{(\kappa k)J} = \sum \frac{h(l_1)}{\Delta E} D(l,l') \begin{Bmatrix} L & 1 & k \\ l & l & l' \end{Bmatrix} \begin{Bmatrix} S & 1 & \kappa \\ \frac{1}{2} & \frac{1}{2} & \frac{1}{2} \end{Bmatrix} \begin{Bmatrix} k & \kappa & J \\ S & L & 1 \end{Bmatrix}$$

$$\times \left(\frac{[1,\kappa,k,S,L]}{[\frac{1}{2},l_1]}\right)^{1/2} (-1)^J [A(l'l_1;l_1l) + A(l_1l';ll_1) - A(l_1l';l_1l)[\frac{1}{2},l_1]]$$

$$\times [(-1)^{S+\kappa} + (-1)^{L+k}]\mathbf{R}^{(\kappa k)J}(nl,nl). \tag{V-22}$$

$$\mathbf{S}^{(\kappa k)J} = \sum \frac{1}{\Delta E} \begin{Bmatrix} S & 1 & \kappa \\ \frac{1}{2} & \frac{1}{2} & \frac{1}{2} \end{Bmatrix} \begin{Bmatrix} L & 1 & k \\ l & l & l \end{Bmatrix} \begin{Bmatrix} \kappa & k & J \\ L & S & 1 \end{Bmatrix} [S,L,k,\kappa,1]^{1/2} D(l'l)$$

$$\times B(l,l')(-1)^J ((-1)^{S+\kappa} + (-1)^{L+k}) \mathbf{W}^{(\kappa k)J}(nl,nl). \tag{V-25}$$

Note. Limits of summation are discussed in the text.

we have assumed that $\mathbf{T}^{(SL)}$ is hermitian, which implies that $D(l,l') = (-1)^{l+l'} D(l',l)$. In Eq. (V-13), l_1 is the filled shell from which the electron is moved; l', the empty shell to which it is moved. The term $2[k]^{-1} C^k(l_1 l_2; l_3 l_4)$ which appears in Eq. (V-13) is defined above and is explicitly given by

$$\frac{2}{[k]} C^k(l_1 l_2; l_3 l_4) = 2e^2 \sum_{\text{all } j} [j_1, j_2, j_3, j_4] \begin{Bmatrix} \frac{1}{2} & \frac{1}{2} & 0 \\ l_1 & l_3 & k \\ j_1 & j_3 & k \end{Bmatrix}$$

$$\times \begin{Bmatrix} \frac{1}{2} & \frac{1}{2} & 0 \\ l_2 & l_4 & k \\ j_2 & j_4 & k \end{Bmatrix} \begin{pmatrix} j_1 & k & j_3 \\ -\frac{1}{2} & 0 & \frac{1}{2} \end{pmatrix} \begin{pmatrix} j_2 & k & j_4 \\ -\frac{1}{2} & 0 & \frac{1}{2} \end{pmatrix} (-1)^{j_1+j_2+1}$$

$$\times \Delta(l_1 l_3 k) \Delta(l_2 l_4 k) \iint \frac{r_<^k}{r_>^{k+1}} (F_1(r_1) F_3(r_1) + G_1(r_1) G_3(r_1))$$

$$\times (F_2(r_2) F_4(r_2) + G_2(r_2) G_4(r_2)) \, dr_1 \, dr_2. \tag{V-14}$$

The pertinent Feynman graphs describing the two terms in Eq. (V-13) are given in Figures V-1a and V-1b, respectively. The perturbation due to the "effective Coulombic" part of U_{cr} vanishes for all S, L, and J of interest. Note that Δ_1 is constant for all states in the configuration C_0.

Let us now consider the effect of the "spin-orbit" part of $(e^2/r_{12})_{\text{trans}}$ in inducing transitions to intermediate configurations C_i of the first kind. In this case, we find that the effect of the perturbation term cannot be simply interpreted as a shielding of $D(l,l)$ as was possible in the previous case. Instead, we must add new operators $\mathbf{N}^{(\kappa k)J}$ to the Hamiltonian to describe the effect of this perturbation. These new operators are given by Eq. (V-15) (Table V-1). In obtaining Eq. (V-15), we have written the "spin-orbit" part of $(e^2/r_{12})_{\text{trans}}$ as

$$\sum_{i \neq j} A(l_1 l_2; l_3 l_4) \, \mathbf{r}_i^{(00)0}(n_1 l_1, n_3 l_3) \cdot \mathbf{r}_j^{(11)0}(n_2 l_2, n_4 l_4); \quad (V\text{-}16)$$

where, using Eqs. (III-34)–(III-36), we find

$$A(l_1 l_2; l_3 l_4) = e^2 \sum_{j_1, j_2} \left(\frac{[1]}{[\tfrac{1}{2}, l_1]} \right)^{\!\!\tfrac{1}{2}} [j_1, j_2] \, \delta(j_1, j_3) \, \delta(j_2, j_4)$$

$$\times \delta(l_1, l_3) \, \delta(l_2, l_4) \begin{Bmatrix} \tfrac{1}{2} & \tfrac{1}{2} & 1 \\ l_2 & l_2 & j_2 \end{Bmatrix} (-1)^{l_2 + j_2 - \tfrac{1}{2}}$$

$$\times \iint \frac{1}{r_>} (F_1(r_1) F_3(r_1) + G_1(r_1) G_3(r_1))$$

$$\times (F_2(r_2) F_4(r_2) + G_2(r_2) G_4(r_4)) \, dr_1 \, dr_2.$$

Clearly, in this expression, although $l_1 = l_3$, n_1 is not necessarily equal to n_3; an equivalent statement can be made for l_2, l_4 and n_2, n_4. In Eq. (V-15), the possible values of κ and k run over all values permitted by the 6-j symbols. The interactions which lead to Eq. (V-15) are all of the type described by the Feynman diagrams of Figure V-1b. The perturbation due to the "spin-orbit" part of U_{cr} vanishes for all S, L, and J of interest.

It is worthwhile to consider this last perturbation term in more detail. We can use Eq. (III-13) to reduce A to a more familiar looking non-relativistic expression:

$$A(l_1 l_2; l_3 l_4) = -\frac{\hbar^2}{2m^2 c^2} [l_2(l_2+1)(2l_2+1)(2l_1+1)]^{\tfrac{1}{2}} \delta(l_2, l_4)$$

$$\times \delta(l_1, l_3) \int R_2(r_2) R_4(r_2) \frac{1}{r_2} \frac{d}{dr_2} U(l_1 l_3) \, dr_2 \quad (V\text{-}17)$$

where

$$U(l_1 l_3) = e^2 \int \frac{R_1(r_1) R_3(r_1)}{r_>} \, dr_1.$$

The non-relativistic limit of the "spin-orbit" part of $(e^2/r_{12})_{\text{trans}}$ is then given by the expression

$$\sum_{i \neq j} A(l_1 l_2; l_3 l_4) \mathbf{w}_i^{(00)0}(n_1 l_1, n_3 l_3) \cdot \mathbf{w}_j^{(11)0}(n_2 l_2, n_4 l_4) \quad (V\text{-}18)$$

where A is defined in Eq. (V-17). Clearly, this operator describes the portion of the spin-orbit interaction which arises from the interaction of the jth

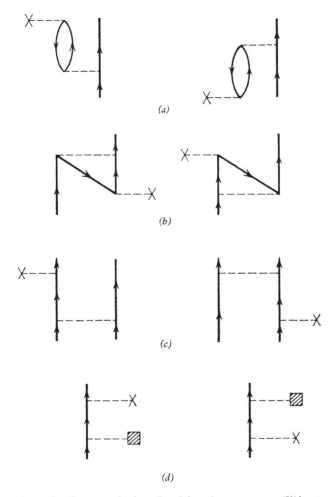

Figure V-1. Second-order perturbations involving the operator $T^{(SL)J}$. The symbol X represents the operator $T^{(SL)J}$; the shaded box represents the central potential.

electron with the gradient of the central field produced by all of the other electrons. Just as clearly, this is a two-body interaction. The usual spin-orbit interaction, written as $\sum a_i \mathbf{s}_i \cdot \mathbf{l}_i$, is simply an effective operator valid for diagonal matrix elements which can be obtained using the Wigner-Eckart theorem in conjunction with the correct expression given above. The fact that the spin-orbit interaction has, in reality, a two-body part has been emphasized in the past,[69] but this fact has generally been ignored.[70-73] In this case, the importance of considering the two-body nature of the interaction is clear: no perturbation at all would be possible in the above circumstances if one considered the spin-orbit interaction to be completely a one-body effect.

b. Excitation of an Electron from a Partially Filled to an Empty Shell

Let us now consider the case in which the electron is moved from a partially filled shell $(n_1 l_1)$ to an empty shell $(n'l')$. Investigating first the effect of the "Coulomb" part of $(e^2/r_{12})_{\text{trans}}$, we find that the perturbation can be calculated within the ground configuration by adding to the Hamiltonian the two-body effective operator $\mathbf{M}^{(SL)J}$ which is given by Eq. (V-19) of Table V-1. In this equation the first sum is over κ, k, l_α, l_β, l_γ, l_ϕ, and l'; the sum over the l_i's runs over all partially filled shells with the restriction that either $l_\alpha = l_\phi$, $l_\beta = l_\gamma$ or $l_\alpha = l_\gamma$, $l_\beta = l_\phi$. The sum over l' runs over all empty shells. The Coulombic part of $-U(r)$ causes a perturbation which can be reproduced through a scaling of $D(l, l)$ by the factor Δ_2 which is given by Eq. (V-20) of Table V-1. In Eq. (V-20), the sum over l_1 is over all partially filled shells (including the l shell), l' over all empty shells. The Feynman graphs for these two interactions are given in Figures V-1c and -1d, respectively.

When we consider the "spin-orbit" part of $(e^2/r_{12})_{\text{trans}}$ we find that the effects it produces by the mixing of intermediate configurations of the type discussed in the previous paragraph can be accounted for by introducing the two-body effective operators $\mathbf{P}^{(\kappa k)J}$ described by Eq. (V-21a), where κ and k can have any value allowed by the 6-j symbols which appear in this equation. Further effective operators which must be introduced to describe this perturbation are given by Eq. (V-21b). Both of these equations can be found in Table V-1.

In both of these two-body terms, l and l_1 are partially filled shells (not necessarily different), and the l' shell is empty. The additional parameters in this case, l_α and l_β, can take on the values l and l_1, with the limitation that $n_\alpha l_\alpha \neq n_\beta l_\beta$ unless $nl = n_1 l_1$. Note that the limits on κ and k in Eq. (V-21a) are the same as appear in Eq. (V-15). There is also a one-body term which arises through the configuration mixing caused by the spin-orbit part of U_{cr}. This one-body term is given by Eq. (V-22) of Table V-1; in this equation the values l, l_1, l' are defined as above. In Eqs. (V-19)–(V-22), ΔE refers to the average energy difference between the ground configuration and the configuration in which an $n_1 l_1$ electron has been replaced by an $n'l'$ electron. Actually, in some cases, there is a partial cancellation between Eqs. (V-21a) and (V-22). Pertinent Feynman diagrams for Eqs. (V-21) and (V-22) are given by Figures V-1c and V-1d respectively.

c. Excitation of an Electron from a Filled to a Partially Filled Shell

Finally, let us consider the case in which an electron is taken from the filled $n'l'$ shell and moved to a partially filled $n_1 l_1$ shell. In this case, we find two-body terms given by equations equivalent to the negative of Eqs. (V-19) and (V-21). There are effects which scale $D(l, l)$ by factors which are formally equivalent to those obtained from Eq. (V-13) and the negative of Eq. (V-20). One-body operators arise which can be obtained from equations equivalent to (V-15); other one-body operators arise which can be obtained by taking the negative of Eq. (V-22).

One final type of perturbation remains; that describing the situation in which an electron is excited from one partially filled shell to another. In this case, one will obtain one- and two-body terms such as those above, and three-body terms such as appear in crystal-field-induced configuration interaction.[73] We will not discuss such terms here, however, as we shall limit our study in the future to atoms having only one partially filled shell. The reader can easily make the extension to the general case using the techniques described above.

5. PERTURBATIONS IN THE NONRELATIVISTIC THEORY

The equations derived above concern perturbations on the relativistic wavefunction. These equations can be carried over to the non-relativistic problem with only minimal effort. First, we note that H_p does not contain the spin-orbit interaction, although this interaction is often of such size that it must be considered. Thus, in the non-relativistic case, we will find it advantageous to investigate perturbations caused by H_p plus the spin-orbit interaction (where the two-body part of the spin-orbit interaction is given by Eq. (V-16)). We therefore consider perturbations due to configuration mixing induced by the operator

$$H_p + \sum_{\substack{l \\ i \neq j}} A(l_1 l_2; l_3 l_4) \mathbf{w}_i^{(00)0}(n_1 l_1, n_3 l_3) \cdot \mathbf{w}_j^{(11)0}(n_2 l_2, n_4 l_4)$$

$$+ \sum_{l, i} (Bl_2, l_4) \mathbf{w}_i^{(11)0}(n_2 l_2, n_4 l_4), \quad \text{(V-23)}$$

where

$$B(l_1, l_2) = -\frac{\hbar^2}{2m^2 c^2} \delta(l_1, l_2) \left[\frac{l_1(l_1 + 1)(2l_1 + 1)}{2} \right]^{1/2} \int \frac{R_{n_1 l_1} R_{n_2 l_2}}{r} \frac{d}{dr} \left(-\frac{Ze^2}{r} \right) dr.$$

The operators describing the perturbations induced by H_p for the various types of excited configurations can be obtained directly from Eqs. (V-13), (V-19) and (V-20) by making the substitution of $\mathbf{w}^{\kappa k}$ for $\mathbf{r}^{\kappa k}$ and using the non-relativistic form of $C^k(l_1 l_2 ; l_3 l_4)$:

$$C^k(l_1 l_2 ; l_3 l_4) = R^k(l_1 l_2 ; l_3 l_4)\langle l_1 \| C^{(k)} \| l_3 \rangle \langle l_2 \| C^{(k)} \| l_4 \rangle, \qquad (V-24)$$

as well as the appropriate non-relativistic expression for $D(l, l')$. The perturbations due to the two-body part of the spin-orbit interaction can likewise be simply obtained by using the non-relativistic A of Eq. (17) in Eqs. (V-15) and (V-21).

In addition, we must consider the perturbation due to the one-body part of the spin-orbit interaction. There is no contribution from this term when the excitation involves the promotion of an electron from a filled to an empty shell. When the excitation involves the promotion of an electron from a partially filled nl shell to an empty $n'l'$ shell, the perturbation can be obtained by evaluating the effective operators $S^{(\kappa k)J}$ given in Eq. (V-25) of Table V-1. If we consider a perturbation in which an electron is moved from a filled $n'l'$ shell to a partially filled nl shell, we obtain simply an equation which is formally the negative of Eq. (V-25).

There are, of course, no contributions from Eq. (V-22) in the non-relativistic case as there is no spin-orbit component of U_c. However, because there is no spin-orbit component to U_c, there is no reason why there cannot be excitation caused by the spin-orbit interaction in which the passive electron is in a filled shell. That is, there will no longer be cancellations such as are represented in the diagrams below Eq. (V-5). The effects of this type of perturbation are easily obtained: a one-body effective operator is added to the Hamiltonian. This operator is simply given by Eq. (V-22), where now the values $n_1 l_1$ run over all closed shells only, and $n'l'$ can run over all closed and empty shells. When the $n'l'$ shell is empty, a factor (-1) must be inserted into (V-22).

6. CORE POLARIZATION—PERTURBATION ANALYSIS

It is convenient to describe by the name "core polarization" corrections to the dipole interaction constants which would disappear if a true Hartree-Fock wavefunction were used to evaluate these constants. Such corrections are those which, in the language of perturbation theory, involve in any order the excitation of a single electron from the ground state through the action of H_p (or H_{pr}). In this section, we shall apply the general results of the previous two sections to the study of core polarization. We shall also discuss other techniques by which core polarization can be analyzed. For simplicity, we assume in our discussion that there is no more than one partially filled shell of electrons present (the nl shell), although there may be any number of closed shells.

a. Contributions to A_S

The dipole interaction produced by an s electron is generally much larger than that produced by any other electron. A rough estimate, based on the results of Chapter VIII, shows that $A(s\,\text{electrons}) \approx 3l^3 A(l\,\text{electrons})$ for electrons in the same atom having the same effective quantum number. Perturbations involving s electrons may, consequently, be rather large, often of the same magnitude as the first order contributions to A from electrons with $l > 0$. The second order s electron effect, although small in comparison with the first order s electron expression, may therefore be a very important contributor to A when the first order s electron effect is zero, that is when all s electrons present are in closed shells. A study of perturbations involving the excitation of an s electron is thus a reasonable point at which to begin our investigation of core polarization.

The only part of $\mathbf{M}_D^{(1)}$ which has non-vanishing matrix elements for s electrons is that with angular dependence $\mathbf{R}^{(10)1}$; it is consequently this term which we would expect to find involved in the largest second order corrections to A. The expressions describing perturbation effects given in the previous two sections indicate that the only second order changes in $D_{10}(nl, nl)$ $R^{(10)1}(nl, nl)$ must come from the promotion of an electron in the $n'l'$ shell to the $n''l'$ shell. In the case of interest to us here

$$D_{10}(nl, nl) = \frac{2\mu_0}{3}[2(2l+1)]^{1/2}\left\langle\frac{\delta(r)}{r^2}\right\rangle_{10}$$

where $\langle\delta(r)/r^2\rangle_{10}$ is given by Eq. (IV-20). The off-diagonal parameter $D_{10}(nl, n'l)$ needed for this investigation is given by

$$D_{10}(nl, n'l) = \frac{2\mu_0}{3}[2(2l+1)]^{1/2}\overline{\left\langle\frac{\delta(r)}{r^2}\right\rangle_{10}}$$

where $\overline{\langle\delta(r)/r^2\rangle_{10}}$ is obtained by replacing $P_{jj'}$ with

$$\int \frac{(F_{nlj}G_{n'lj'} + G_{nlj}F_{n'lj'})}{r^2}\, dr$$

in the definition of $\langle\delta(r)/r^2\rangle_{10}$.

If we assume that there are no electrons of the type $n'l'$ in the partially filled shell, then the effect of the perturbation can be described by a Δ from Eq. (V-13).

$$\Delta'_S = -\sum \frac{2D_{10}(n''l', n'l')(-1)^k}{\Delta E D_{10}(nl, nl)[l, l']^{\frac{1}{2}}} C^k(n''l', nl; nl, n'l'). \quad \text{(V-26)}$$

This effect can be taken into account by replacing $\langle \delta(r)/r^2 \rangle_{10}$ by

$$\left\langle \frac{\delta(r)}{r^2} \right\rangle_S = (1 + \Delta'_S) \left\langle \frac{\delta(r)}{r^2} \right\rangle_{10}$$

in Eq. (IV-20). In the non-relativistic limit

$$\left\langle \frac{\delta(r)}{r^2} \right\rangle_S = -\frac{8\pi}{[l]} \sum_{n'',n'} \frac{(n''s|\delta(\mathbf{r})|n's)}{\Delta E} R^l(n''s, nl; nl, n's) \quad \text{(V-27)}$$

when $l > 0$.

In Chapter IX, we demonstrate that $D_{10}(nl, nl)$, l not equal to zero, is of the order of $(Z\alpha)^2 l^{-5} D_{10}(ns, ns)$. Thus, the value of Δ'_S will be produced primarily by excitation of s electrons from the core into empty shells. For large Z, however, one might expect the effect due to promotion of a p electron from the core into an empty shell to become comparable in size to the s electron effect; perturbations due to promotion of electrons with $l > 1$ can be safely ignored insofar as they affect D_{10}.

It is sometimes convenient to write, instead of Δ'_S, a parameter ΔA which represents the change in the dipole interaction constant A_S

$$\Delta A = \sum_{\substack{k \\ n',n'' \\ l'}} \frac{(-1)^k 2 D(n''l', n'l')\mu_I}{\Delta E[l, l']^{\frac{1}{2}} IJ} C^k(n''l', nl; nl, n'l')$$

$$\times (JJ|R^{(10)\,1}_{\,\,0}(nl, nl)|JJ). \quad \text{(V-28a)}$$

This becomes, in the non-relativistic limit[74]

$$\Delta A = (g_J + 1) \sum_{n',n''} \frac{32\pi\mu_I\mu_0}{3 \Delta E[l] I} (n''s|\delta(\mathbf{r})|n's) R^l(n''s, nl; nl, n's). \quad \text{(V-28b)}$$

In arriving at the last result, we have used

$$\mathbf{S} = \mathbf{L} + 2\mathbf{S} - \mathbf{J} = -g_J\mathbf{J} - \mathbf{J}.$$

(See Section VI-1 for the definition of g_J.) This change in A is often considered to be an s electron effect exclusively. It is well to emphasize, however, that this interpretation is correct only for $Z \ll 1/\alpha$, and that for heavy elements p electron core polarization can become important.

Let us now consider the case in which an electron is moved from the partially filled nl shell to an empty shell. One can easily show using the results of Section V-4 that all perturbations to A_S vanish in this case when $l = 0$. Following the arguments given above, we can conclude that only if $l = 1$ is there likely to be an observable effect produced by a perturbation affecting $D_{10}(nl, nl)\mathbf{R}^{(10)1}$, and then only for atoms with large Z. As this type of perturbation is, therefore, not of general interest, let us simply consider the form the perturbation will take. First, Eq. (V-19) simplifies somewhat, with κ being forced equal to k, and k taking on the values 0 and 2. The term in the expansion of $\mathbf{M}^{(10)1}$ having $k = 0$ will behave as a one-body operator with angular dependence $\mathbf{r}^{(10)1}$; the effect of this term will then be simply to scale $D_{10}(p, p)$. The term having $k = 2$ can be shown, using the Wigner-Eckart theorem, to provide a scaling for $D_{10}(p, p)$ which depends on the S and L values of the states used to evaluate matrix elements. That is, this part of $\mathbf{M}^{(10)1}$ provides a Δ_S'' which is constant for all the levels in an SL term, but which varies from term to term.

There are, in addition, a number of perturbations to $\langle \delta(r)/r^2 \rangle_{10}$ which are caused by the second-order interaction between $D_{01}\mathbf{R}^{(01)1}$ and the "spin-orbit" terms of $(e^2/r_{12})_{\text{trans}}$. Many of these perturbations contribute directly to Δ_S': the perturbations of this type are special cases of Eq. (V-15), (V-21a), (V-22) and (V-25). In addition, there is a two-body perturbation of the type described by Eq. (V-21b). The effect of this operator is to provide a scaling Δ_S'' of $\langle \delta(r)/r^2 \rangle_{10}$ which is constant for the levels of a term, but which may vary from term to term. The non-relativistic forms of all of these perturbations are given explicitly in Table V-2, where they are expressed as contributions to $\langle \delta(r)/r^2 \rangle_S$; with $\langle \delta(r)/r^2 \rangle_S'$ those contributions which are constant for all levels of the configuration, and $\langle \delta(r)/r^2 \rangle_S''$, those which vary from term to term.

It is clear from the preceding discussion that the interactions which most affect D_{10} involve the mixing into the ground configuration of an excited state containing an unpaired s (or $p_{1/2}$) electron. This unpaired s (or $p_{1/2}$) electron density at the origin can cause hyperfine anomalies of the Bohr-Weisskopf type (Section VII-2). Far from being a mere annoyance, this attribute is quite helpful in separating core polarization from relativistic effects.[24] As we saw in Section IV-6, in a non-relativistic atom with no s electrons outside of closed shells, the contribution of A from the Hamiltonian term transforming like $\mathbf{S}$ should vanish. If experimentally it is seen not to vanish, one cannot in general tell whether this is due to relativistic interactions or core polarization. However, the relativistic interaction does not imply a non-vanishing spin density at the origin, as does core polarization of the type described by Eqs. (V-26)–(V-28). Thus, if a hyperfine anomaly is measured in an atom having no unpaired s or p electrons, it must signify that at least part of A_S results from core polarization of this type.

Table V-2. Contributions to $\langle \delta(r)/r^2 \rangle_S$

---

a. Contributions to $\langle \delta(r)/r^2 \rangle'_S$

(1) $\quad -\dfrac{8\pi}{[l]} \sum\limits_{\substack{n' \text{ filled} \\ n'' \text{ empty}}} \dfrac{(n''s|\delta(\mathbf{r})|n's)}{\Delta E} R^l(n''s, nl; nl, n's).$ (V-13)

(2) $\quad \dfrac{(l(l+1))^{1/2}}{[l]} \sum\limits_{\substack{n_1 \text{ filled} \\ n' \text{ not filled}}} \dfrac{(n_1 l | 1/r^3 | n'l)}{\Delta E} [A(n'l, nl; nl, n_1 l) + A(nl, n'l; n_1 l, nl)].$ (V-15)

(3) $\quad -(N_l - 1) \dfrac{(l(l+1))^{1/2}}{[l]} \sum\limits_{n' \neq n} \dfrac{(nl|1/r^3|n'l)}{\Delta E} (-1)^S A(nl, n'l; nl, nl).$ (V-21a)

(4) $\quad \left(\dfrac{l(l+1)}{(2l+1)}\right)^{1/2} \sum\limits_{\substack{n_1 l_1 \text{ filled} \\ n' \neq n}} \dfrac{(nl|1/r^3|n'l)}{\Delta E (2l_1 + 1)^{1/2}} (-1)^S [A(n'l, n_1 l_1; n_1 l_1, nl)$

$\qquad + A(n_1 l_1, n'l; nl, n_1 l_1) - [\tfrac{1}{2}, l_1] A(n_1 l_1, n'l; n_1 l_1, nl)].$ (V-22)

(5) $\quad -\left(\dfrac{2l(l+1)}{[l]}\right)^{1/2} \sum\limits_{n' \neq n} (-1)^S \dfrac{(nl|1/r^3|n'l)}{\Delta E} B(nl, n'l).$ (V-25)

b. Contributions to $\langle \delta(r)/r^2 \rangle''_S$

(1) $\quad \left(\dfrac{2l(l+1)}{3[l]}\right)^{1/2} \sum\limits_{n' \neq n} \dfrac{(nl|1/r^3|n'l)}{\Delta E} A(n'l, nl; nl, nl)(-1)^S$

$\qquad \times \dfrac{\langle SL \| \sum_{i \neq j} (\mathbf{w}_i^{(01)} \mathbf{w}_j^{(11)})^{(10)} \| SL \rangle}{\langle SL \| W^{(10)} \| SL \rangle}.$ (V-21b)

Note. Results quoted in this table include the effects discussed in Sections V-4a, V-4b, and V-4c. The factor S is equal to one if $n'l'$ is a filled shell, zero if $n'l'$ is not filled. The number following each entry refers to the general equation of which the entry is a special case.

b. Contributions to A_L

Most of the remaining perturbations can also be considered in their non-relativistic limits without loss of pertinent information. For simplicity of presentation, we continue to assume that all electrons outside of closed shells are in the nl shell. The term in the dipole interaction with angular dependence **L** can induce only perturbations of the type $n'l' \to n''l'$. If neither $n'l'$ nor $n''l'$ is equal to nl the perturbation resulting from the interaction of this term with the Coulombic term can simply be taken into account by multiplying the expectation value of $1/r^3$ which appears in A_L by the factor $(1 + \Delta'_L)$, where (Eq. (V-13))[45,75]

$$\Delta'_L = \sum_{\substack{k,n',n'' \\ l'}} \frac{2}{\Delta E} \left[\frac{l'(l'+1)(2l'+1)}{l(l+1)(2l+1)} \right]^{1/2} \frac{(n''l'|1/r^3|n'l')}{(nl|1/r^3|nl)} \begin{Bmatrix} l & 1 & l \\ l' & k & l' \end{Bmatrix}$$
$$\times \langle l' \| C^{(k)} \| l \rangle^2 R^k(n''l', nl; nl, n'l'). \quad \text{(V-29)}$$

If, on the other hand, one considers the promotion of an nl electron to an $n'l$ shell, the hyperfine term with angular dependence **L** must be augmented with a two-body interaction because of the coulombic perturbations (Eq. (V-19)):

$$\mathbf{M}^{(01)1} = \frac{8\mu_0}{3} \sum_{\kappa,k,n'} \frac{1}{\Delta E} \left(\frac{[\kappa]}{[k]} \right)^{1/2} [3l(l+1)(2l+1)]^{1/2} (n'l|1/r^3|nl) \langle l \| C^{(k)} \| l \rangle^2$$
$$\times R^k(nl, nl; n'l, nl) \begin{Bmatrix} 1 & \kappa & k \\ l & l & l \end{Bmatrix} ((-1)^\kappa - 1)$$
$$\times \sum_{i<j} (\mathbf{w}_i^{(0\kappa)}(nl, nl) \mathbf{w}_j^{(0k)}(nl, nl))^{(01)1}. \quad \text{(V-30)}$$

Again using the Wigner-Eckart theorem, one can conclude that the effect of such an operator will be to shield (or enhance) $\langle 1/r^3 \rangle$ by some fraction Δ''_L of $\langle 1/r^3 \rangle$, where Δ''_L will be constant for all levels of a term, but will vary from term to term. Eq. (V-30) is expressed in terms of Δ''_L in Table V-3. In this case, there will also be a scaling of $\langle 1/r^3 \rangle$ caused by the interaction of the central field with the dipole term proportional to **L** (Eq. (V-20)), which is constant for all levels in the configuration. The explicit form of this contribution to Δ'_L is given in Table V-3. Similar results to those given by (V-29) and (V-30) occur for excitations from a filled $n'l$ shell to the nl shell. These contributions to Δ_L are also indicated in Table V-3.

There are also a number of contributions to Δ_L produced by the interaction between $D_{12} \mathbf{W}^{(12)1}$ and the spin-orbit operator of Eq. (V-18). The perturbations described by Eqs. (V-15), (V-21a), (V-22), and (V-25) contribute to Δ'_L; those described by Eq. (V-21b) contribute to Δ''_L. We have written out the explicit forms of these contributions to Δ_L in Table V-3.

c. Contributions to A_{sC}

We now consider perturbations which can effect A_{sC}. Looking first at perturbations which do not involve promotion of electrons from or to the nl shell, we find that the interaction involving $D_{12} \mathbf{W}^{(12)1}$ and the Coulomb term

Table V-3. Contributions to Δ_L

a. Contributions to Δ_L'

(1) $\displaystyle\sum_{\substack{k,n_1l' \text{ filled} \\ n'l' \text{ not filled}}} \frac{2}{\Delta E} \left[\frac{l'(l'+1)(2l'+1)}{l(l+1)(2l+1)}\right]^{1/2} \frac{(n_1l'|1/r^3|n'l')}{(nl|1/r^3|nl)} \langle l'\|C^{(k)}\|l\rangle^2$

$\times \begin{Bmatrix} l & 1 & l \\ l' & k & l' \end{Bmatrix} R^k(n_1l,'nl;nl,n'l').$ (V-13)

(2) $\displaystyle\frac{-1}{2[l](l(l+1))^{1/2}} \sum_{\substack{n_1 \text{ filled} \\ n' \text{ not filled}}} \frac{1}{\Delta E} \frac{(n_1l|1/r^3|n'l)}{(nl|1/r^3|nl)} [A(n'l,nl;nl,n_1l)$

$+ A(nl,n'l;n_1l,nl)].$ (V-15)

(3) $\displaystyle\frac{N_l}{[l]} \sum_{n' \neq n} (-1)^S \frac{1}{\Delta E} \frac{(nl|1/r^3|n'l)}{(nl|1/r^3|nl)} \{\langle l\|C^{(0)}\|l\rangle^2 R^0(n'l,nl;nl,nl)$

$- \sum_k \langle l\|C^{(k)}\|l\rangle^2 [l]^{-1} R^k(n'l,nl;nl,nl)\}.$ (V-20)

(4) $\displaystyle\frac{N_l-1}{2[l](l(l+1))^{1/2}} \sum_{n' \neq n} \frac{(-1)^S}{\Delta E} A(nl,n'l;nl,nl) \frac{(nl|1/r^3|n'l)}{(nl|1/r^3|nl)}.$ (V-21a)

(5) $\displaystyle\frac{-1}{2(l(l+1)(2l+1))^{1/2}} \sum_{\substack{n_1l_1 \text{ filled} \\ n' \neq n}} \frac{(-1)^S}{\Delta E[l_1]^{1/2}} \frac{(nl|1/r^3|n'l)}{(nl|1/r^3|nl)}$

$\times [A(n'l,n_1l_1;n_1l_1,nl) + A(n_1l_1,n'l;nl,n_1l_1) - [\tfrac{1}{2},l_1]A(n_1l_1,n'l;n_1l_1,nl)].$ (V-22)

(6) $\displaystyle\frac{1}{(2l(l+1)(2l+1))^{1/2}} \sum_{n' \neq n} \frac{(-1)^S}{\Delta E} B(nl,n'l) \frac{(n'l|1/r^3|nl)}{(nl|1/r^3|nl)}.$ (V-25)

b. Contributions to Δ_L''

(1) $\displaystyle 4 \sum_{\substack{\kappa,k, \\ n' \neq n}} \frac{(-1)^S}{\Delta E} \left(\frac{[\kappa]}{[\tfrac{1}{2},k]}\right)^{1/2} \frac{(n'l|1/r^3|nl)}{(nl|1/r^3|nl)} \langle l\|C^{(k)}\|l\rangle^2$

$\times R^k(nl,nl;n'l,nl) \begin{Bmatrix} 1 & \kappa & k \\ l & l & l \end{Bmatrix} ((-1)^\kappa - 1)$

$\times \frac{\langle SL\|\sum_{i<j}(\mathbf{w}_i^{(0\kappa)}\mathbf{w}_j^{(0k)})^{(01)}\|SL\rangle}{\langle SL\|W^{(01)}\|SL\rangle}.$ (V-19)

(2) $\displaystyle 2 \sum_{n' \neq n} \frac{(-1)^S}{\Delta E} \left[\frac{1}{3(2l+3)(2l-1)(2l+1)}\right]^{1/2} \frac{(nl|1/r^3|n'l)}{(nl|1/r^3|nl)}$

$\times A(n'l,nl;nl,nl) \frac{\langle SL\|\sum_{i<j}(\mathbf{w}_i^{(12)}\mathbf{w}_j^{(11)})^{(01)}\|SL\rangle}{\langle SL\|W^{(01)}\|SL\rangle}.$ (V-21b)

Note. Results quoted in this table include the effects discussed in Sections V-4a, V-4b, and V-4c. The factor S is equal to one if $n'l'$ is a filled shell, zero if $n'l'$ is not filled. The number following each entry refers to the general equation of which the entry is a special case.

can be accounted for by multiplying $\langle 1/r^3 \rangle$ of A_{sC} by $(1 + \Delta'_{sC})$, where, from Eq. (V-13),[45]

$$\Delta'_{sC} = \sum_{\substack{n'l' \\ n''l'' \\ k}} \frac{2}{\Delta E} \frac{\langle l''\|C^{(2)}\|l'\rangle}{\langle l\|C^{(2)}\|l\rangle} \frac{(n''l''|1/r^3|n'l')}{(nl|1/r^3|nl)} (-1)^{k+1} \begin{Bmatrix} l & 2 & l \\ l' & k & l'' \end{Bmatrix}$$

$$\times \langle l''\|C^{(k)}\|l\rangle\langle l\|C^{(k)}\|l'\rangle R^k(n''l'', nl; nl, n'l'). \quad \text{(V-31)}$$

Equivalent perturbations involving the promotion of an electron from the partially filled nl shell introduce a two-body operator into the hyperfine Hamiltonian (Eq. (V-19)):

$$\mathbf{M}^{(12)1} = -2\sqrt{2}\mu_0 \sum_{\substack{n'l' \\ \kappa, k}} \frac{1}{\Delta E} \left(\frac{[\kappa]}{[k]}\right)^{1/2} \langle l'\|C^{(2)}\|l\rangle (n'l'|\frac{1}{r^3}|nl)$$

$$\times \langle l\|C^{(k)}\|l\rangle\langle l\|C^{(k)}\|l'\rangle R^k(nl, nl; n'l', nl)$$

$$\times \begin{Bmatrix} 2 & k & \kappa \\ 1 & l & l' \end{Bmatrix} ((-1)^\kappa + 1) \sum_{i<j} (\mathbf{w}_i^{(1\kappa)}(nl, nl)\mathbf{w}_j^{(0k)}(nl, nl))^{(12)1}. \quad \text{(V-32)}$$

Again, the effect of this term is to introduce a scaling of the $\langle 1/r^3 \rangle$ of A_{sC} by a factor Δ''_{sC} which is constant for all levels of a term, but which varies from term to term. Equation (V-32) is expressed in terms of Δ''_{sC} in Table V-4. There is also a contribution to Δ'_{sC} from the interaction of $D_{12}\mathbf{W}^{(12)1}$ and the central field; this term can also be found in Table V-4. Excitations from a filled $n'l'$ shell to the nl shell produce results similar to these; Table V-4 also contains such contributions.

There are, in addition, terms which result from the interaction between the spin-orbit interaction and the operators $D_{01}\mathbf{W}^{(01)1}$ and $D_{12}\mathbf{W}^{(12)1}$. A number take the form of a one-body interaction: these are special cases of Eqs. (V-15), (V-20), (V-21a), (V-22), and (V-25), and contribute to Δ'_{sC}. There is also a two-body term which is a special case of Eq. (V-21b); this interaction contributes to Δ''_{sC}. Explicit forms for all of the above contributions are given in Table V-4.

Collecting the results of this section, we find that, when core polarization is taken into account, Eq. (IV-36) becomes

$$H_D = \frac{2\mu_I\mu_0}{I} \sum_i \left(\frac{2}{3}\left\langle\frac{\delta(r)}{r^2}\right\rangle_s \mathbf{s}_i + \left\langle\frac{1}{r^3}\right\rangle_L \mathbf{l}_i - (10)^{1/2}\left\langle\frac{1}{r^3}\right\rangle_{sC} (\mathbf{s}_i\mathbf{C}_i^{(2)})^{(1)}\right) \cdot \mathbf{I}$$

where

104   *The Central Field Approximation*

$$\left\langle \frac{\delta(r)}{r^2} \right\rangle_S = \left\langle \frac{\delta(r)}{r^2} \right\rangle'_S + \left\langle \frac{\delta(r)}{r^2} \right\rangle''_S$$

$$\left\langle \frac{1}{r^3} \right\rangle_L = (1 + \Delta'_L + \Delta''_L)\left\langle \frac{1}{r^3} \right\rangle = (1 + \Delta_L)\left\langle \frac{1}{r^3} \right\rangle \quad \text{(V-33)}$$

$$\left\langle \frac{1}{r^3} \right\rangle_{sC} = (1 + \Delta'_{sC} + \Delta''_{sC})\left\langle \frac{1}{r^3} \right\rangle = (1 + \Delta_{sC})\left\langle \frac{1}{r^3} \right\rangle.$$

The radial parameters $\langle \delta(r)/r^2 \rangle_S$, $\langle 1/r^3 \rangle_L$, and $\langle 1/r^3 \rangle_{sC}$ are constant within a term, but may vary from term to term. Thus, because of configuration interaction effects, the $\langle 1/r^3 \rangle$ parameters of A_L and A_{sC} may differ even in the absence of relativistic corrections. On the other hand, relativistic effects can only cause the parameter $\langle 1/r^3 \rangle_{01}$ to differ from $\langle 1/r^3 \rangle_{12}$ by a fraction

Table V-4. Contributions to Δ_{sC}

---

a. Contributions to Δ'_{sC}

(1) $\displaystyle\sum_{\substack{n_1 l_1 \text{ filled} \\ n'l' \text{ not filled} \\ k}} \frac{2}{\Delta E} \frac{\langle l' \| C^{(2)} \| l_1 \rangle}{\langle l \| C^{(2)} \| l \rangle} \frac{(n'l' | 1/r^3 | n_1 l_1)}{(nl | 1/r^3 | nl)} (-1)^{k+1}$

$\times \begin{Bmatrix} l & 2 & l \\ l_1 & k & l' \end{Bmatrix} \langle l' \| C^{(k)} \| l \rangle \langle l \| C^{(k)} \| l_1 \rangle R^k(n'l', nl; nl, n_1 l_1).$ (V-13)

(2) $\displaystyle -\frac{(8l^2 + 8l + 3)}{6[l](l(l+1))^{1/2}} \sum_{\substack{n_1 \text{ filled} \\ n' \text{ not filled}}} \frac{1}{\Delta E} \frac{(n_1 l | 1/r^3 | n'l)}{(nl | 1/r^3 | nl)}$

$\times [A(n'l, nl; nl, n_1 l) + A(nl, n'l; n_1 l, nl)].$ (V-15)

(3) $\displaystyle \frac{N_l}{[l]} \sum_{n' \neq n} \frac{(-1)^S}{\Delta E} \frac{(nl | 1/r^3 | n'l)}{(nl | 1/r^3 | nl)} \{\langle l \| C^{(0)} \| l \rangle^2 R^0(n'l, nl; nl, nl)$

$- \sum_k \langle l \| C^{(k)} \| l \rangle^2 [l]^{-1} R^k(n'l, nl; nl, nl)\}.$ (V-20)

(4) $\displaystyle \frac{(N_l - 1)(8l^2 + 8l + 3)}{6[l](l(l+1))^{1/2}} \sum_{n' \neq n} \frac{(-1)^S}{\Delta E} \frac{(nl | 1/r^3 | n'l)}{(nl | 1/r^3 | nl)} A(nl, n'l; nl, nl).$ (V-21a)

(5) $\displaystyle -\frac{(8l^2 + 8l + 3)}{6(l(l+1)(2l+1))^{1/2}} \sum_{\substack{n_1 l_1 \text{ filled} \\ n' \neq n}} \frac{(-1)^S}{\Delta E [l_1]^{1/2}} \{A(n'l, n_1 l_1; n_1 l_1, nl)$

$+ A(n_1 l_1, n'l; nl, n_1 l_1) - [\tfrac{1}{2}, l_1] A(n_1 l_1, n'l; n_1 l_1, nl)\} \dfrac{(nl | 1/r^3 | n'l)}{(nl | 1/r^3 | nl)}.$ (V-22)

(6) $\displaystyle \frac{(8l^2 + 8l + 3)}{3(2l(l+1)(2l+1))^{1/2}} \sum_{n' \neq n} \frac{(-1)^S}{\Delta E} \frac{(n'l | 1/r^3 | nl)}{(nl | 1/r^3 | nl)} B(nl, n'l).$ (V-25)

Table V-4 (*Continued*)

b. Contributions to Δ''_{sc}

(1) $\quad 2\sqrt{2} \sum\limits_{\substack{\kappa, k \\ n'l' \neq nl}} \dfrac{(-1)^S}{\Delta E} \left(\dfrac{[\kappa]}{[k]}\right)^{1/2} \dfrac{(n'l'|1/r^3|nl)}{(nl|1/r^3|nl)} \dfrac{\langle l'\|C^{(2)}\|l\rangle}{\langle l\|C^{(2)}\|l\rangle} \langle l\|C^{(k)}\|l'\rangle$

$\qquad \times \langle l\|C^{(k)}\|l\rangle R^k(nl, nl; n'l', nl) \begin{Bmatrix} 2 & k & \kappa \\ l & l & l' \end{Bmatrix}$

$\qquad \times ((-1)^\kappa + 1) \dfrac{\langle SL\|\sum_{i<j}(\mathbf{w}_i^{(1\kappa)}\mathbf{w}_j^{(0k)})^{(12)}\|SL\rangle}{\langle SL\|W^{(12)}\|SL\rangle}.$ (V-19)

(2) $\quad \dfrac{4}{[\tfrac{1}{2}, l]^{1/2}} \sum\limits_{n' \neq n} \dfrac{(-1)^S}{\Delta E} \dfrac{(nl|1/r^3|n'l)}{(nl|1/r^3|nl)} A(n'l, nl; nl, nl)$

$\qquad \times \left\{ \dfrac{1}{3} \left[\dfrac{10(2l+3)(2l-1)}{3}\right]^{1/2} \dfrac{\langle SL\|\sum_{i<j}(\mathbf{w}_i^{(01)}\mathbf{w}_j^{(11)})^{(12)}\|SL\rangle}{\langle SL\|W^{(12)}\|SL\rangle} \right.$

$\qquad \left. + \dfrac{1}{2} \dfrac{\langle SL\|\sum_{i<j}(\mathbf{w}_i^{(12)}\mathbf{w}_j^{(11)})^{(12)}\|SL\rangle}{\langle SL\|W^{(12)}\|SL\rangle} \right\}.$ (V-21b)

Note. Results quoted in this table include the effects discussed in Sections V-4a, V-4b, and V-4c. The factor S is equal to one if $n'l'$ is a filled shell, zero if $n'l'$ is not filled. The number following each entry refers to the general equation of which the entry is a special case.

which is constant for all levels in a configuration; promotion of an electron from, or to, a partially filled shell results in a scaling of $\langle 1/r^3\rangle_L$ and $\langle 1/r^3\rangle_{sC}$ which is constant within the levels of a term, but which varies from term to term. Thus, if it is experimentally found that the $\langle 1/r^3\rangle$'s of A_{sC} and A_L scale differently from term to term, core polarization involving promotion of an electron from, or to, a partially filled shell must be involved.

It is worth noting that thus far we have considered only the interaction between the Coulomb and spin-orbit terms and those parts of $\mathbf{M}^{(1)}$ which have parallel parts in $\mathbf{M}_D^{(1)}$. There will also be interactions between the remaining parts of $\mathbf{M}^{(1)}$, that is, a part transforming like $\mathbf{R}^{(11)1}$, and the Coulomb and spin-orbit terms. These remaining interactions are, however, completely relativistic, and disappear in the non-relativistic limit. We can expect such interactions, therefore, to be very small, and we shall not consider them further.

The sums appearing in the above expressions can be carried out by calculating excited state wavefunctions and evaluating each term in the sum. Although a difficult calculation to carry out because of the number of terms in the sum which must be considered, this calculation does demonstrate the importance of the different parts of the sum. Judd[75] has carried out such a calculation for oxygen ($1s^2 2s^2 2p^4$) which is worth describing in some detail. The central potential of Herman and Skillman[76] was used to determine the needed radial wavefunctions. The calculation was carried out for an atom in the ^3P state; the change in $\langle 1/r^3 \rangle$ for A_{sc} due to promotion of a $2p$ electron to higher p states (Eq. (V-19)) was found to be $0.035 \langle 1/r^3 \rangle$, with the contribution to this number from bound states being only about one third of the total. The change in $\langle 1/r^3 \rangle$ for A_{sc} due to promotion of core $1s$ and $2s$ electrons to excited d states (Eq. (V-13)) is of the same order, $0.025 \langle 1/r^3 \rangle$. In this instance, the entire change is due to promotion of the s electron to d-like continuum states. This analysis clearly indicates not only the importance of interactions with the continuum, but also the importance of excitation of the type $n'l' \to n''l''$, where $l' \neq l''$. This type of excitation is often called angular, where excitations of the type $n'l' \to n''l'$ are called radial. Bauche and Judd[74] hypothesized that the continuum is of such importance in calculations of this sort because the intermediate states needed nodal properties quite unlike those of the unexcited states. If this hypothesis has general validity, it would imply that the perturbations of Section V-4c, which do not involve the continuum, should generally be much smaller than the perturbations of Sections V-4a and V-4b, which do involve the continuum.

In many cases, the equations given above can be used without carrying out the indicated summations. For example, Bordarier, Judd, and Klapisch[77] have used such equations in a study of s electron core polarization in EuI $4f^7 6s6p$, EuI $4f^7 6s^2$, EuII $4f^7 6s$, and EuIII $4f^7$. In the states of interest, the electrons in the $4f$ shell could be considered to be in the $^8S_{7/2}$ state; in this case, the first order hyperfine interaction arises entirely from the $6p$ electron and the contact interaction of the $6s$ electron (if relativity is neglected). There are, in second order, core polarization effects proportional to the total spin of the $4f$ electrons, the total spin of the $6s$ electrons, and the spin of the $6p$ electron (these effects are similar to those described by Eq. (V-28)). Bordarier, et. al.[77] found that all of these types of polarization in all the configurations of interest can be expressed in terms of only six parameters if one assumes that all s electron eigenfunctions, all f electron eigenfunctions, and all excitation energies are equivalent for these different electronic structures. These parameters are of the same general form as Eq. (V-28b), with, for example, A_1 obtained by setting $nl = 4f$, $ns = 6s$, and carrying out a sum over $n'' > 6$ in Eq. (V-28b). The results of Bordarier, et. al. are shown in Table V-5; core polarization produced by the l electrons is described by the operator $\delta a_l S_l$. One can use the results of Table V-5 to relate various core polarization parameters to experimentally determined parameters. For example, the entire measured A values in EuI $4f^7 6s^2\,^8S$ and EuIII $4f^7\,^8S$ arise solely from δa_f; the measured hyperfine constant for EuI $4f^7 6s^2\,^8S$

Table V-5. Core Polarization[77] in Eu

Configuration	Type of Excitation ($n' > 6 > n$)	δa_f	δa_p	δa_s
EuI $4f^76s^2$	$(ns)^2(6s)^2 \to (ns)^2(6s)(n's)$	A_1	—	—
	$(ns)^2(6s)^2 \to (ns)(6s)^2(n's)$	A_2	—	—
EuI $4f^76s6p$	$(ns)^2(6s) \to (ns)^2(n's)$	$\tfrac{1}{2}A_1$	$\tfrac{1}{2}A_3$	$(7/2)A_1 + \tfrac{1}{2}A_3$
	$(ns)^2(6s) \to (ns)(6s)(n's)$	A_2	A_4	0
	$(ns)^2(6s) \to (ns)(6s)^2$	$\tfrac{1}{2}A_5$	$\tfrac{1}{2}A_6$	$-(7/2)A_5 - \tfrac{1}{2}A_6$
EuII $4f^76s$	$(ns)^2(6s) \to (ns)^2(n's)$	$\tfrac{1}{2}A_1$	—	$(7/2)A_1$
	$(ns)^2(6s) \to (ns)(6s)(n's)$	A_2	—	0
	$(ns)^2(6s) \to (ns)(6s)^2$	$\tfrac{1}{2}A_5$	—	$-(7/2)A_5$
EuIII $4f^7$	$(ns)^2 \to (ns)(6s)$	A_5	—	—
	$(ns)^2 \to (ns)(n's)$	A_2	—	—

($-.688$ mK) is thus equal to $A_1 + A_2$, and the corresponding constant for EuIII $4f^7\,^8S$ (-3.430 mK) is equal to $A_2 + A_5$. Measured p electron core polarization further indicates that A_3, A_4, and A_6 may be small. If such is the case, one can easily obtain the contribution of δa_s to the hyperfine structure of EuIII $4f^76s6p$:

$$\delta a_s(4f^76s6p) = \tfrac{7}{2}A_1 + \tfrac{1}{2}A_3 - \tfrac{7}{2}A_5 - \tfrac{1}{2}A_6$$
$$\approx \tfrac{7}{2}A_1 - \tfrac{7}{2}A_5$$
$$= \tfrac{7}{2}\delta a_f(4f^76s^2) - \tfrac{7}{2}\delta a_f(4f^7)$$
$$= 9.67 \text{ mK}.$$

7. CORE POLARIZATION—HARTREE-FOCK TREATMENT

The defects in the central field model which give rise to the effects discussed above can also be handled within the framework of the Hartree-Fock formalism by improving the restricted Hartree-Fock.

The exact details of the calculations are beyond the scope of this work, and, in any case, there is still a considerable diversity of opinion in the literature as to the best way to carry out the calculations. Two general approaches have been suggested which have considerable appeal. In the first, the configuration interaction technique,[78-82] the ground state used as a basis for the Hartree-Fock calculation is composed not of a single configuration, but of the nominal "ground state" configuration and one or more configurations differing from the "ground" configuration by one or two electrons. The energy is then minimized with respect to this combination of configurations using usual Hartree-Fock techniques. Of course, one cannot consider all possible admixtures into the "ground" configuration, as this would lead to a problem of insurmountable complexity. One must therefore obtain an "average" form for the excited admixtures, a not altogether satisfactory procedure. This is especially true in light of the results of Judd[75] for oxygen, which appear to indicate the need of excited states possessing many different nodal properties. A second approach is known as the spin polarized Hartree-Fock, and its less often used variation, the orbital polarized Hartree-Fock.[80-87] In the former case, the restriction the R_{nl} be independent of m_s is relaxed, with electrons with spin up satisfying a radial equation different from those with spin down. In the latter case, this freedom is extended to electrons of differing m_l. Because different states of the same nl have different radial wavefunctions in this approximation, the expectation value of, for example, $\sum_i (\mathbf{s}_i \mathbf{C}_i^{(2)})^{(1)}/r_i^3$ evaluated for a closed shell need no longer vanish. Thus there may be a closed shell contribution to the hyperfine interaction when polarized Hartree-Fock wavefunctions are used. The relationship of the polarized Hartree-Fock methods to the Configuration Interaction Methods is discussed by Bessis et al.[79] Obviously, the polarized Hartree-Fock type of calculation takes into account only radial excitations, since it does not change the angular nature of the wavefunctions. However, s electron core polarization can be treated accurately using spin polarized Hartree-Fock wavefunctions because this effect involves only a radial excitation of electrons with $l = 0$. Consequently, spin polarized Hartree-Fock calculations are the type most often encountered in the literature. An unfortunate consequence of allowing R_{nl} to depend on m_s is that the wavefunction ceases to be an eigenfunction of $\mathbf{S}^2$. Whether this does, or does not, affect the accuracy to which hyperfine interactions can be evaluated is open to discussion. One can project out the part of the wavefunction belonging to a particular eigenvalue of $\mathbf{S}^2$; this should be done, ideally, before the Hartree-Fock minimization process is carried out.[88] This procedure, unfortunately, introduces major complications into the calculation and has never actually been done. (See, however, Goddard.[89])

We must emphasize that the techniques described above are not the only ones which can be used to study perturbations to the magnetic dipole interaction. For example, Sternheimer[90] used perturbation theory to derive equations which expressed the radial wavefunction of the perturbed state in terms of the radial wavefunction of the unperturbed state. The technique used in this case is much like that which we shall describe in the next section. Gaspari, et al.[91] worked in the Hartree-Fock scheme, but approached the problem from a different direction than that followed by the authors mentioned above. Gaspari, et al. assumed the core electron wavefunctions to be perturbed by the nuclear magnetic moment; these perturbed wavefunctions were then used for the Hartree-Fock minimization process. More recently, many-body techniques have been applied to the evaluation of hyperfine interactions. These techniques, which will be discussed in detail in Section V-9, have the added value of allowing correlation to be included in the calculation. Although many body techniques would appear, at least in principle, to be the best method to use in the calculation of hyperfine constants, the very great complexity of these calculations would seem to indicate a continued reliance on the more straightforward polarized Hartree-Fock technique for heavy atoms and ions.

Although the calculations are not trivial, when sufficient data exist, one can extract from experimental results values of ΔA or $\langle \delta(r)/r^2 \rangle_s$ by considering the dependence of the effect on S. Abragam et al.[92] consider such an effect in $3d^N$ neutral atoms and divalent ions (Table V-6). Childs[93] has carried out a more recent analysis for $3d^N$ atoms (Table V-7). For heavier atoms, relativistic effects having the same S dependence as core polarization must be taken into account: see, for example the analyses of Pu by Bauche and Judd,[74] Armstrong[94] and Lewis et al.[95] The spin polarized Hartree-Fock method predicts quite well the observed effect for $3d^N$ divalent ions.[85] An

Table V-6. $\left\langle \dfrac{\delta(r)}{r^2} \right\rangle_s$ for $3d^N$ Ions

	Experimental[92] (au)	Theoretical[84] (au)
$V^{2+} 3d^3$	−2.8	
$Mn^{2+} 3d^5$	−3.1	−3.34
$Fe^{3+} 3d^5$		−3.00
$Fe^{2+} 3d^6$		−3.29
$Co^{2+} 3d^7$	−2.5	
$Ni^{2+} 3d^8$		−3.94
$Cu^{2+} 3d^9$	−2.9	

Table V-7. $\left\langle \dfrac{\delta(r)}{r^2} \right\rangle_s$ for $3d^N$ Neutral Atoms

	Experimental[93] (au)	Theoretical[84,85] (au)
Sc $3d^1$	−0.28	−0.17
Ti $3d^2$	−0.53	
V $3d^3$	−0.67	−0.45
Mn $3d^5$	−0.83	−0.54
Fe $3d^6$	−0.90	
Co $3d^7$	−0.97	−0.60
Ni $3d^8$	−1.27	
Cu $3d^9$	−1.92	−0.67

interesting result of these calculations[84,92] is that the core polarization is roughly independent of $Z(\langle\delta(r)/r^2\rangle_S \approx -3$ au. Table V-6). The calculations for $3d^N$ atoms are much more difficult to carry out due to cancellations between shells, but the results[85] are still in reasonable agreement with experiment ($\langle\delta(r)/r^2\rangle_S \approx -0.5$ au. Table V-7). Fewer experimental results are known for $4d$ atoms and ions; calculations[85,86] show that the core polarization effects are about three times as large for $4d$ ions as for $3d$ divalent ions ($\langle\delta(r)/r^2\rangle_S \approx -8.5$ au) and much more independent of not only Z but also the degree of ionization. Again with $4f$ electrons, there is a paucity of experimentally determined core polarization parameters. Here, calculations[85] suggest that core polarization should be roughly independent of Z for triply ionized rare earths, and only one-half to one-third as large as the value for divalent $3d^N$ ions ($\langle\delta(r)/r^2\rangle_S \approx -1.2$ au). Bleaney[96] has suggested that core polarization in neutral rare earths should be very small, a prediction which appears to be supported by recent calculations in Eu^{95} and $Sm.^{95,97}$ These calculations[95] also suggest that core polarization in the neutral actinides may be very small.

When sufficient data have been collected, one can also extract from experimental results the different values of $\langle 1/r^3 \rangle$ which appear in A_L and A_{sC}. As examples of such calculations, one has the work of Childs on the $3d^N$ atoms[93]; Childs and Goodman[98] on cobalt and vanadium; Radford and Evenson[99] on nitrogen; and Woodgate[100] on samarium. (All of these works include, of course, a determination of ΔA.) C. Bauche-Arnoult[101] has analyzed the results of Childs[93] using the parameterization technique of Bordarier et al.[77]; by doing so, she has been able to explain some of the variation observed in the $\langle 1/r^3 \rangle_{\text{eff}}$ values of the $3d^N$ series. Some of the most studied results, however, are those obtained by Harvey[102] for oxygen and fluorine; he found that, for both atoms, $\langle 1/r^3 \rangle$ of A_{sC} must be approximately 10% larger than $\langle 1/r^3 \rangle$ of A_L. These results have been studied with varying degrees of success by, for example, Judd[75] using second-order perturbation theory, Bessis, et al.[81] and Schaefer, et al.[80] using configuration interaction techniques, and Kelly[103] using many-body techniques.

8. QUADRUPOLE SHIELDING

The quadrupole interaction constant b is affected by the same types of mechanisms as discussed in the previous sections. Again, we find it convenient to first consider the problem from the standpoint of perturbation theory. The equations of Section V-4 and V-5 can be used for this purpose by replacing $\mathbf{T}^{SL}$ by the quadrupole operators of Eq. (IV-21). In Section IV-6 we showed that the terms with angular dependence $\mathbf{R}^{11}$ and $\mathbf{R}^{13}$ will in general be small; we would therefore expect second-order effects due to these terms to be small enough to ignore. In any case, no studies of the second order effects of these terms have been carried out. The remaining term transforming like $\mathbf{R}^{02}$ will, for simplicity, be considered in the non-relativistic limit and it is assumed that all electrons outside of closed shells are in the nl shell.

The effect on the quadrupole interaction of raising an electron via the Coulomb interaction from a filled $n'l'$ shell to the open $n''l''$ shell can be absorbed into H_Q or b by multiplying the expectation value of $1/r^3$ by $(1 + \Delta_Q)$, where, from Eq. (V-13)[45, 72]

$$\Delta_Q = \sum_{\substack{n'l' \text{ filled} \\ n''l'' \text{ not filled}}} \frac{2}{\Delta E} \frac{\langle l''\|C^{(2)}\|l'\rangle \langle n''l''|1/r^3|n'l'\rangle}{\langle l\|C^{(2)}\|l\rangle \langle nl|1/r^3|nl\rangle} \{ \tfrac{2}{5} \langle l\|C^{(2)}\|l\rangle$$

$$\times \langle l''\|C^{(2)}\|l'\rangle R^2(nl, n''l''; nl, n'l') + \sum_k (-1)^{k+1} \begin{Bmatrix} l & 2 & l \\ l' & k & l'' \end{Bmatrix}$$

$$\times \langle l''\|C^{(k)}\|l\rangle \langle l\|C^{(k)}\|l'\rangle R^k(n''l'', nl; nl, n'l') \}. \tag{V-34}$$

This scaling effect is, as in the dipole case, constant for all states in the configuration nl^N. A two-body operator results from the promotion via the Coulomb interaction of an nl electron into an empty shell $n'l'$ or from a filled $n'l'$ shell into the nl shell (Eq. (V-19))

$$2\sqrt{2} \sum_{\substack{\kappa, k \\ n'l' \neq nl}} \frac{1}{\Delta E} \left(\frac{[\kappa]}{[2, k]}\right)^{1/2} \langle l'\|C^{(2)}\|l\rangle (n'l'|1/r^3|nl)\langle l\|C^{(k)}\|l'\rangle$$

$$\times \langle l\|C^{(k)}\|l\rangle R^k(nl, nl; n'l', nl) \begin{Bmatrix} 2 & k & \kappa \\ l & l & l' \end{Bmatrix} (-1)^S$$

$$\times ((-1)^\kappa + 1) \sum_{i<j} (\mathbf{w}_i^{(0\kappa)} \mathbf{w}_j^{(0k)})^{(02)2}. \tag{V-35}$$

The factor S is equal to one if the $n'l'$ shell is filled, and is equal to zero when this $n'l'$ shell is not filled. This term can be absorbed into $\langle 1/r^3 \rangle$ by assuming a different effective $\langle 1/r^3 \rangle$ for each SL term. It is customary to define a parameter R which expresses the scaling of $\langle 1/r^3 \rangle$ for a given SL term due to both the one and two body effects above; according to this definition, quadrupole shielding is taken into account if $\langle 1/r^3 \rangle$ of H_Q is replaced by $\langle 1/r^3 \rangle (1 - R) = \langle 1/r^3 \rangle_Q$. Occasionally one finds in the literature an R which is the negative of this R; the reader should take care to ascertain for himself which definition prevails when he reads an article on this subject.

There are also perturbations to b resulting from spin-orbit induced configuration mixing. These interactions produce one-body effective operators (Eqs. (V-15), (V-21a), (V-22)) with angular dependence $\mathbf{W}^{(11)2}(nl, nl)$ and $\mathbf{W}^{(13)2}(nl, nl)$. There is, of course, also a two-body operator with angular dependence $\sum_{i \neq j} (\mathbf{w}_i^{(02)}(nl, nl)\mathbf{w}_j^{(11)}(nl, nl))$. Therefore, if it is found that operators with angular dependence $\mathbf{W}^{(11)2}$ or $\mathbf{W}^{(13)2}$ are required to properly explain experimental data, one can conclude that either relativistic effects or second-order spin-orbit interactions are present. Of course, since the latter is relativistic in origin, it is perhaps true that when the spin-orbit interaction is large enough to produce second-order effects, the whole problem should be considered only from a relativistic standpoint.

Wolter[104] has studied shielding effects in boron, oxygen, aluminum and manganese using equations similar to those above. Results of this calculation are qualitatively quite similar to the results of Judd[75] for the equivalent dipole calculation (Section V-6). For example, in the case of oxygen, Wolter finds that excitations of the type $2p - p'$ make up approximately two-thirds of the scaling of $\langle 1/r^3 \rangle$, with the bound $2p - p'$ excitation contributing only about one-half as much as the continuum $2p - p'$ excitations. In the case of $ns - d$ excitations, virtually the entire contribution is attributable to excitations into d-like continuum states. Similar results are found for other atoms with, in general, angular excitations contributing positively to R and radial excitations contributing negatively to R. We speak of the former excitation as "shielding" the quadrupole moment (lessening the magnitude of the gradient of the electric field at the nucleus), and the latter as "antishielding" the quadrupole moment.

The study of quadrupole shielding was originated by Sternheimer and he has continued to be one of the principle contributors to the literature of the field.[105-110] Thus, it is not surprising that quadrupole shielding factors are often referred to as "Sternheimer factors." His approach is, like the one above, based on perturbation theory. In this case, however, perturbation theory is used to evaluate the radial portion of excited states. As is obvious from the second order perturbation expression at the beginning of Section V-4, effects of the type we are considering can be evaluated by perturbing a wavefunction with the Coulomb interaction and then evaluating the quadrupole interaction using these perturbed wavefunctions, or, alternatively, by perturbing the wavefunction with the quadrupole interaction and then evaluating the Coulomb interaction with these perturbed wavefunctions. The Sternheimer approach is perhaps clearest if we follow the latter procedure. We start with the equation

$$H_0 \psi_0 = E_0 \psi_0$$

where H_0 is a Hamiltonian with some type of central field, that is expressable as a sum of one electron Hamiltonians. We perturb these wavefunctions with the quadrupole interaction

$$H_Q = -\frac{e^2}{2} \sum_i \frac{1}{r_i^3} C_{i0}^{(2)} Q. \tag{IV-39}$$

The perturbed wavefunction can be written as $\psi_0 + \psi_1$, the perturbed energy as $E_0 + E_1$. Ordinary perturbation theory immediately gives

$$(H_0 - E_0)\psi_1 = -(H_Q - E_1)\psi_0. \tag{V-36}$$

Let us assume that $\psi_0 = \psi_\alpha \ldots \psi_\varepsilon \ldots \psi_\delta$ where the ψ_α are single electron wavefunctions, $\alpha = (nlm_s m_l)$. This, of course, ignores antisymmetrization and is analogous to a Hartree calculation of the central potential. Then, because H_0 and H_Q are composed of sums of one electron operators, we can write $\psi_1 = \psi_\alpha \ldots \psi_\phi \ldots \psi_\delta$ where the replacement $\psi_\varepsilon \to \psi_\phi$ is the only change between ψ_0 and ψ_1. With wavefunctions of this type, Eq. (V-36) simplifies

$$(H_0 - E_0)_i \psi_\phi = -(H_Q - E_1)_i \psi_\varepsilon$$

where the subscripts i indicate the use of the Hamiltonian and energy of the ith electron only. In detail, this becomes

$$\left(-\frac{\hbar^2}{2m}\frac{1}{r}\frac{d^2 r}{dr^2} + \frac{l^2 \hbar^2}{2mr^2} + V(r_i) - E_i\right)\psi_\phi = \frac{e^2 Q}{2}\left[\frac{C_0^{(2)}}{r^3} - \langle\psi_\varepsilon|\frac{C_0^{(2)}}{r^3}|\psi_\varepsilon\rangle\right]\psi_\varepsilon.$$

We can expect ψ_ϕ to be given by a sum of wavefunctions of differing orbital momentum:

$$\psi_\phi = \sum_a C_a \psi_a.$$

If we further separate the ψ_a into radial and angular parts

$$\psi_a = Y_{l_a m_a}(\theta, \phi)\, u_{l_\varepsilon \to l_a}(r)/r$$

and assume that $C_a = (\psi_a|C_0^{(2)}|\psi_\varepsilon)$, the above equation leads to the equations for the $u_{l_\varepsilon \to l_a}$

$$\left(-\frac{\hbar^2}{2m}\frac{d^2}{dr^2} + \frac{l_a(l_a+1)\hbar^2}{2mr^2} + V(r) - E_i\right)u_{l_\varepsilon \to l_a}$$

$$= \frac{e^2 Q}{2}\left[\frac{1}{r^3} - \left\langle\frac{1}{r^3}\right\rangle_\varepsilon \delta(l_\varepsilon, l_a)\right]u_\varepsilon. \tag{V-37}$$

Sternheimer solves the above equations for $u_{l_\varepsilon \to l_a}$ by using Hartree or Hartree-Fock wavefunctions for u_ε. The perturbed wavefunctions $\psi_0 + \psi_1$ are then used to evaluate e^2/r_{ij}; part of the expectation value of e^2/r_{ij} will be of the form of an additional interaction between the nucleus and the atomic electrons, that is, proportional to Q. This interaction, when compared with the usual quadrupole interaction arising from the open shell electrons only, gives immediately the shielding due to distortions of the closed, but not completely spherical, shells. Sternheimer finds that, in general, the radial excitations lead to antishielding, the angular excitations to shielding. If the valence electron is outside of the core, the main excitation is usually radial, leading to the conclusion that antishielding usually dominates in excited states. For ground states, Sternheimer has found that both the direct and exchange parts of the Coulomb interaction are of great importance; for excited states the direct often predominates. R is generally found to be of the order $|R| < 0.3$.

As in the case of dipole interaction, quadrupole shielding can be considered in the framework of the Hartree-Fock theory by using either the polarized Hartree-Fock or the configuration interaction approach. A serious limitation of the former approach is that, with the usual Hartree-Fock formalism, only radial excitations can be considered. This is a limitation which often makes the calculations of little value for low lying states, where angular excitations may predominate. The configuration interaction technique suffers, as in the dipole case, from the very large number of excited states which might possibly be of importance. On the other hand, a great advantage of the Hartree-Fock type of calculations is that they automatically take into account many higher order perturbations. For example, not only the distortion of the electronic shell by the valence electrons is considered by these procedures, but also the second order effect of the distorted core on itself and on the valence shell; Sternheimer generally considers only the former effect.

In Table V-8 are listed a number of references in which values of R have been presented. We have listed only "average" values because the details of the calculations carried out by any one author may differ from paper to paper, and the techniques used by different authors may differ greatly. Therefore, it seems imperative that the reader, before using any of these calculated R's, acquaint himself with the particular approximations used in the particular case.

Table V-8. Calculated Values of R

Z	State	Approximated R	Reference
3	$2p$	0.12	106, 110
	$3p$	0.10	110
	$4p$	0.10	108a
4	$2s2p\,^3P$	0.04	110
5	$2p$	0.04	80, 81, 104, 105, 106, 107, 127a
6	^3P	0.04	80
8	^1D	0.02	104
	^3P	0.1	80, 81, 103, 104
9	$2p^5$	0.1	80, 106
11	$3p$	-0.2	107, 108
	$4p$	-0.14	108a
	$5p$	-0.12	108a
13	$3p$	-0.2	104, 105, 106

Table V-8 (*Continued*)

17	$3p^5$	-0.2	106, 107, 108, 112
19	$4p$	-0.2	108, 108a
	$5p$	-0.17	108, 108a
	$6p$	-0.16	108a
21	$3d$	-0.2	105, 106
25	$3d^54s4p^6P$	-0.29	104
	$3d^54s4p^8P$	-0.27	104
26	$(Fe^{2+})3d^6\,^5D$	0.2	111, 112
29	$3d^94s^2$	0.18	106, 107, 108, 110
	$3d^{10}4p$	-0.18	110
31	$4p$	-0.03	105, 106
35	$4p^5$	-0.02	106
37	$4p$	0.03	107
	$5p$	-0.25	107, 108, 108a
	$6p$	-0.2	108, 108a
	$7p$	-0.17	108, 108a
39	$4d$	-0.1	106
47	$4d$	-0.08	106
49	$5p$	-0.02	105, 106
53	$5p$	-0.02	106
55	$5p$	-0.02	107
	$6p$	-0.24	107, 108, 108a
	$7p$	-0.21	108, 108a
	$8p$	-0.19	108a
57	$5d$	-0.11	106
58	$(Ce^{3+})4f$	0.3	113
59	$5d$	-0.38	110
	$(Pr^{3+})4f^2$	0.15	109, 114
63		-0.2	105, 106
69	$5d$	-0.44	110
	$(Tm^{3+})4f^{12}$	0.13	109, 114
71	$4f$	-0.15	106
74	$5d^4$	-0.5	107
78	$5d$	-0.05	106
81	$6p$	-0.01	106
85	$6p$	-0.01	106
89	$6d$	-0.06	105, 106
92	$5f$	-0.15	106

Experimental determination of R is somewhat more difficult than the experimental determination of the dipole scaling factors. In the dipole case, A_{sC}, A_S and A_L contribute to A in different proportions for different levels. Values of A measured in different levels can be expressed in terms of equations involving $\langle 1/r^3 \rangle_L$, $\langle 1/r^3 \rangle_{sC}$, and $\langle \delta(r)/r^2 \rangle_s$; if A has been measured in more than three levels, these radial parameters can be determined by solution of simultaneous equations. However, in the quadrupole interaction all b values in a configuration (at least, those arising from levels of the same term) will be scaled by the same fraction. Thus one must usually compare different configurations in order to experimentally obtain numbers for R.

Zu Putlitz and co-workers[115] have managed to carry out such a comparison in both cesium and rubidium. They first measured b in a series of $np_{3/2}$ states. A value of Q was then extracted from each b using equations such as those of Chapter VIII and $\langle 1/r^3 \rangle$ values obtained from the fine structure splitting of the np states. These Q's, however, are not equal to the real quadrupole moment of the nucleus, because $\langle 1/r^3 \rangle_Q$, not the real $\langle 1/r^3 \rangle$, must be used to extract Q from b. Thus the measured Q's are equal to $1 - R$ times the real Q, and ratios of Q's measured in different np states must be equal to ratios of $1 - R$ for those same np states. The ratios determined by Zu Putlitz, et al.[115] for cesium and rubidium are in almost complete agreement with predictions of Sternheimer. To some extent, the completeness of the agreement must be considered accidental because of the uncertainty in the determination and application of the "real" $\langle 1/r^3 \rangle$ values obtained from the fine structure. However, even with this uncertainty, these results must be considered to be a powerful confirmation of Sternheimer's predictions.

Another impressive confirmation of Sternheimer's results is to be found in the determination of the quadrupole moment of Cu^{65}. The values of Q derived from the hyperfine structures of the $3d^9 4s^2$ and the $3d^{10} 4p$ configurations differ by about 40% when Sternheimer corrections are disregarded, but are almost exactly equal when these corrections are included.[110] Additional experimental determinations of R have been carried out by, for example, Murakawa and co-workers.[116]

9. CORRELATION

In the event that a true Hartree-Fock wavefunction can be obtained for the atomic state of interest, the perturbations discussed in the previous two sections would vanish. There are, however, perturbations of a different type which would still have to be considered. These arise from correlation effects.[117-128]

Correlation energy is defined as the difference between the measured energy of an electron (with the relativistic contribution to the energy subtracted out) and the Hartree-Fock energy of the electron. As was demonstrated in Section V-2, there are no second-order shifts in the Hartree-Fock energy due to excitation of a single electron from the ground state; there are, however, shifts caused by excitation of pairs of electrons from this ground state. In third, and higher, order perturbation theory, these pair excitations can influence hyperfine structure.

Many techniques have been developed to study correlation effects. The configuration interaction modification of the Hartree-Fock procedure discussed in Section V-7 can be used; this method is often very slowly converging.[117] Sinanoglu[118] and Nesbet[119] use equations which allow the N-body problem to be divided into $N(N-1)/2$ two-body problems. For example, Nesbet studies two particle excitations by using the Hamiltonian for particles i and j

$$H_{ij} = \frac{p_i^2}{2m} + \frac{p_j^2}{2m} - \frac{Ze^2}{r_i} - \frac{Ze^2}{r_j} + (i_1 j_2 | \frac{e^2}{r_{12}} | i_1 j_2)$$

$$- (i_1 j_2 | \frac{e^2}{r_{12}} | j_1 i_2) + \sum_{\alpha \neq j} \left[(i_1 \alpha_2 | \frac{e^2}{r_{12}} | i_1 \alpha_2) - (i_1 \alpha_2 | \frac{e^2}{r_{12}} | \alpha_1 i_2) \right]$$

$$+ \sum_{\alpha \neq i} \left[(j_1 \alpha_2 | \frac{e^2}{r_{12}} | j_1 \alpha_2) - (j_1 \alpha_2 | \frac{e^2}{r_{12}} | \alpha_1 j_2) \right]. \quad \text{(V-38)}$$

The purpose of this calculation is to study excitations which remove electrons i and j from their Hartree-Fock states but which leave all of the remaining electrons unperturbed; the sum over the states $\alpha \neq i, j$ above can therefore be carried out using Hartree-Fock states. Nesbet then expands the eigenfunction of H_{ij} as a sum of the Hartree-Fock wavefunction plus all possible determinantal product states in which only electrons i and j differ from Hartree-Fock values:

$$\psi_{ij} = \Phi_{HF} + \sum_{a,b} \Phi_{ab}^{ij} c_{ab}^{ij} + \sum_a \Phi_a^i c_a^i + \sum_b \Phi_b^j c_b^j$$

where Φ_{ab}^{ij} is Φ_{HF} with the states i and j replaced by the states a and b, etc. If a true Hartree-Fock wavefunction is used, the coefficients c_a^i and c_b^j can be expected to be small as they are non-vanishing only in second and higher order of perturbation theory. However, since a true Hartree-Fock is generally not used, these coefficients may be of some size. A variational technique is then employed to minimize the eigenvalue E_{ij} of H_{ij} with respect to the coefficients c_{ab}^{ij}, c_a^i, and c_b^j. Three-body Hamiltonians analagous to Eq. (V-38) can also be set up and solved, but the three-body correlations are probably insignificant.[119]

Kelly[103,120] and Sandars[121] have considered the application of many-body diagram techniques to the calculation of hyperfine effects. Kelly calculates correlation energies using a Brueckner-Goldstone[122] linked cluster expansion of the evolution operator of time dependent perturbation theory. In this method, one considers the perturbation H_p of Eq. (V-2) to be turned on slowly, reaching full strength at $t = 0$; then

$$\psi(t) = U(t, -\infty)\phi_0$$

where ϕ_0 is an eigenfunction of the central field Hamiltonian H_0. The evolution operator $U(t, -\infty)$ can be expressed as

$$U(t, -\infty) = \sum_{N=0}^{\infty} (-i)^N \int_{0<t_1<t_2\cdots<t_N} H_p(t_1) H_p(t_2) \cdots H_p(t_N)\, dt_1\, dt_2 \cdots dt_N$$

where $H_p(t) = e^{iH_0 t} H_p e^{\alpha t} e^{-iH_0 t}$. Here, α is a number related to how quickly the perturbation is turned on. For $t = 0$, we have simply

$$\psi(0) = U(0, -\infty)\phi_0.$$

Using Wick's theorem,[123] $U(0, -\infty)\phi_0$ can be expanded as a sum of terms, each of which may be represented by a Feynman graph. Kelly has developed techniques for summing types of related graphs which appear in the hyperfine calculations, and for carrying out sums over all possible excited states.

In the work by Nesbet,[119] the use of restricted Hartree-Fock wavefunctions leads to sizable coefficients $c_a{}^i$. In the expansion used by Kelly,[103,120] use of restricted Hartree-Fock wavefunctions assures that graphs containing interactions with passive ground state electrons will not exactly cancel the diagrams containing the interaction with the central potential (Section V-2). Thus, expectation values of hyperfine interactions calculated using wavefunctions obtained with these techniques will contain corrections due both to inaccuracies in the restricted Hartree-Fock method and to correlation effects. It is convenient to define[120] as "generalized core polarization effects" those perturbations which would vanish, and as "correlation effects" those perturbations which would remain, if a true Hartree-Fock wavefunction were used to calculate the hyperfine constants. In the notation of Kelly,[103,120] contributions to core polarization come from all terms represented by Feynman graphs containing one or more interactions involving a passive unexcited state or the central potential $U(r)$. All other graphs represent correlation effects. Thus

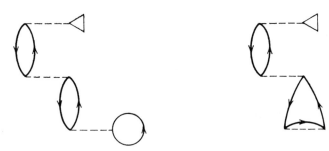

represent third-order core polarization terms, and

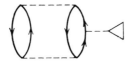

represents a third-order correlation term.

One characteristic of the Brueckner-Goldstone method which makes it very useful is the ease with which its results may be compared with results obtained from other methods. For example, Chang, et al.,[124] have discussed the relationship of different terms in this expansion to various forms of restricted Hartree-Fock calculations. Of course, no restricted Hartree-Fock calculation evaluates the correlation contribution to hyperfine structure, so it is of great interest to determine the importance of correlation in hyperfine structure. It is also pertinent to consider what fraction of the total core polarization arises from the first-order term discussed in the preceding two sections.

Results of various Brueckner-Goldstone calculations are given in Tables V-9 through V-11. The perturbations to A_L and A_{sC} are written in terms of the Δ functions of Section V-8. We see that in oxygen, the contact term is very poorly predicted by the second order term alone.[103] However, because of the lack of contribution from correlation effects the contact interaction should be correctly predicted using Hartree-Fock wavefunctions. In fact, excellent agreement with the experimental result has been obtained by Bessis, et al.[81] using a projected-unrestricted-Hartree-Fock approximation to the correct Hartree-Fock wavefunctions. Continuing with oxygen, we see that for Δ_L the second order term is about two thirds of the total core polarization contribution, with correlation contributing about 12% of the total value of Δ_L. For Δ_{sC}, the second order term provides about 90% of the total core polarization contribution; the effect of correlation in this case is to cancel almost one third of the polarization term. For Δ_Q, the second order term makes up about 80% of the total due to core polarization; the effect of correlation is to add a term equal to roughly 10% of the core polarization. As we saw in Section V-6, angular excitations are important for calculations of this type in oxygen; therefore, polarized Hartree-Fock wavefunctions cannot be expected to be particularly successful in predicting even the core polarization contributions to the various Δ's. In the ground state of lithium, roughly 15% of the contact interaction is a product of correlation; almost 75% of the core polarization contribution is a product of the second order term.[125] In phosphorus for the first time we see a situation in which the correlation effect becomes dominant in the contact interaction.[126] In this case, the correlation contribution is twice as large as the total core polarization, and of opposite sign. It is clear that only the techniques of this section

Table V-9. Calculated Values of A_s (Mc/sec)

Element	Second-Order Core Polarization	Higher-Order Core Polarization	Correlation	Normalization	Total Calculated	Experimental	Reference
O^{17}	−4.18	−19.71	−0.36	6.15	−18.1	−17.2	103
P^{31}	−64.6	12.3	102.0	—	49.7	55.0	126
$Li^7(^2S)$	70.7	23.9	16.2	4.1	114.9	116.6[a]	125
$Li^7(^2P)$	−8.6	−2.0	1.0	—	−9.6	—	127
N	5.8	6.2	−1.4	−0.15	10.5	10.4	128

[a] Obtained by subtracting the contribution of the 2s electron,[125] $A(2s) = 285.1$ Mc/sec, from the measured value, $A = 401.8$ Mc/sec.

Table V-10. Calculated[103] Values of Δ for $O^{17}(^3P)$

Parameter	Second-Order Core Polarization	Higher-Order Core Polarization	Correlation	Normalization	Total Calculated	Experimental[a]
Δ_L	−0.050	−0.029	−0.011	0.004	−0.086	−0.079
Δ_{sc}	0.043	−0.004	−0.012	0.003	0.030	0.043
Δ_Q	−0.123	−0.026	−0.015	0.010	0.154	—

[a] Experimental values obtained using Harvey's[102] $\langle 1/r^3 \rangle_L$ and $\langle 1/r^3 \rangle_{sc}$ and Kelley's[103] value of $\langle 1/r^3 \rangle = 4.975$ au.

Table V-11. Calculated[127] Values of Δ for $Li^7(^2P)$

Parameter	Second Order Core Polarization	Higher Order Core Polarization	Correlation	Total Calculated	Experimental
Δ_L	0	−0.001	0.073	0.072	—
Δ_{sc}	0.108	0.001	0.063	0.172	—
Δ_Q	−0.156	0.004	−0.018	−0.170	—

are sufficient for the calculation of the spin contact interaction in phosphorus. Correlation is also seen to provide essentially all of the Δ_L of Li^2P, and more than one third of Δ_{sc}.[127]

These results demonstrate that both higher-than-second-order core polarization and correlation may be very important in hyperfine structure calculations. Unfortunately, because of the very few calculations which have been made thus far, no rules can be suggested at this time for estimating the general importance of these high-order effects.

VI. Hyperfine Structure in an External Field

1. EXTERNAL MAGNETIC FIELDS

Thus far in our discussion, we have assumed that the atoms were moving in a space with no preferred direction In this chapter we shall consider the effects of externally applied fields on the atoms. One effect of such a perturbation is to destroy the spherical symmetry of the surroundings of the atom, and to define a preferred direction in space. When the external field is uniform, this means that the highest group that can describe the symmetry operations of the new Hamiltonian is $R(2)$, the rotation group in a plane perpendicular to the external field. Thus I, J, and F will cease to be exact quantum numbers, with M_I, M_J and M_F becoming the new good quantum numbers. In general, however, we shall consider only fields that are small enough that I and J will be approximately good quantum numbers.

We have seen that the nucleus, which we have considered to be a non-relativistic body, may possess a magnetic dipole moment $\mathbf{N}^{(1)}$. If the nucleus does possess such a moment, then the interaction between this dipole moment and an external magnetic field is given simply by the classical expression

$$-\mathbf{N}^{(1)} \cdot \mathbf{H}$$

where $\mathbf{H}$ is the applied external field. It is traditional to define the nuclear Lande g factor by

$$g_I = \frac{\mu_I}{\mu_0 I}$$

such that the interaction of the nucleus with an external magnetic field is given by

$$-g_I \mu_0 \mathbf{I} \cdot \mathbf{H}. \tag{VI-1}$$

The effect of the external magnetic field on the electrons is more complicated since we treat the electrons relativistically. A constant magnetic field produces a vector potential given by

$$\mathbf{A} = -\tfrac{1}{2}(\mathbf{r} \times \mathbf{H}).$$

The change in the Dirac Hamiltonian for the ith electron due to such a vector will be

$$e\boldsymbol{\alpha} \cdot \mathbf{A} = \frac{i\sqrt{2}}{2} er(\boldsymbol{\alpha} C^{(1)})^{(1)} \cdot \mathbf{H}. \tag{VI-2}$$

Summing over all of the electrons in the atom, and using the Wigner-Eckart theorem to relate matrix elements of $(\boldsymbol{\alpha} C^{(1)})^{(1)}$ to matrix elements of $\mathbf{J}$ within a set of states arising from a given J, we find the interaction between the electrons and the external field can be expressed as

$$-g_J \mu_0 \mathbf{J} \cdot \mathbf{H}. \tag{VI-3}$$

The Landé g factor, g_J, is given by

$$g_J = -\frac{ie\sqrt{2}}{\mu_0 2J} (JJ| \sum_i r_i(\boldsymbol{\alpha}_i C_i^{(1)})_0^{(1)} |JJ). \tag{VI-4a}$$

We note that definitions of g_J and g_I may differ; other common definitions differ by a sign from the ones given above. The operator whose expectation value is g_J, which we shall call $\mathbf{g}_J$, is simply

$$\mathbf{g}_J = -\frac{i\sqrt{2}}{2} \frac{er}{J\mu_0} \sum_i (\boldsymbol{\alpha}_i C_i^{(1)})^{(1)} = -\frac{r^3}{2\mu_0 J} \mathbf{M}_D^{(1)}$$

or

$$g_J = -\frac{1}{2\mu_0 J} (JJ|r^3 M_{D0}^{(1)}|JJ). \tag{VI-4b}$$

Thus, $\mathbf{g}_J$ can be obtained from Eq. (IV-20) by replacing the integrals $P_{jj'}$ with the integrals $D_{jj'}$, where

$$D_{jj'} = -\frac{1}{2\mu_0 J} \int r(F_{nlj} G_{nlj'} + F_{nlj'} G_{nlj}) \, dr.$$

Taking non-relativistic limits as before, we find

$$D_{++} = -\frac{(2l+3)}{2eJ}$$

$$D_{+-} = -\frac{1}{eJ}$$

$$D_{--} = \frac{2l-1}{2eJ}$$

such that in the non-relativistic limit

$$\mathbf{g}_J = -(\mathbf{L} + 2\mathbf{S})\frac{1}{J}. \qquad \text{(VI-5)}$$

We shall not discuss here in any greater detail the effects of relativistic and other corrections (e.g. Schwinger correction) to the non-relativistic expression for g_J. This subject has been discussed in some detail by several authors.[129]

With these results, the hyperfine interaction in the presence of an external magnetic field can be written as

$$H = A\mathbf{I} \cdot \mathbf{J} + b\,\frac{3(\mathbf{I} \cdot \mathbf{J})^2 + 3/2(\mathbf{I} \cdot \mathbf{J}) - I(I+1)J(J+1)}{2IJ(2I-1)(2J-1)}$$
$$+ c\{10(\mathbf{I} \cdot \mathbf{J})^3 + 20(\mathbf{I} \cdot \mathbf{J})^2 + 2(\mathbf{I} \cdot \mathbf{J})[-3I(I+1)J(J+1) + I(I+1)$$
$$+ J(J+1) + 3]$$
$$- 5I(I+1)J(J+1)\}[(I)(I-1)(2I-1)(J)(J-1)(2J-1)]^{-1}$$
$$- g_J \mu_0 \mathbf{J} \cdot \mathbf{H} - g_I \mu_0 \mathbf{I} \cdot \mathbf{H}. \qquad \text{(VI-6)}$$

All of the terms in this expression except the last two commute with $\mathbf{F}$; the last two commute with J_z and I_z respectively, where the external field is applied along the z axis. In a low magnetic field, the first three terms will dominate, and we can still consider our basis states to be the states $|IJFM_F\rangle$, treating the last two terms as first order perturbations. In a high magnetic field, the last two terms will be largest, so our basis states should be those which diagonalize those terms, that is $|IM_I JM_J\rangle$, and the first three terms should then be treated as first order perturbations. (At very high magnetic fields, $|JM_J\rangle$ states break up into $|LM_L SM_S\rangle$ states. Hyperfine measurements are generally made at fields low enough that we need not consider that effect here.) In the low field limit, the first three terms can be expressed conveniently through the number

$$K = 2(\mathbf{I} \cdot \mathbf{J}) = F(F+1) - J(J+1) - I(I+1).$$

That is, eigenvalues of the first three terms in the hyperfine interaction can be obtained in the weak field case by replacing $\mathbf{I} \cdot \mathbf{J}$ in every term by $K/2$. The Wigner-Eckart theorem is called upon again to evaluate the field-dependent terms. When F is a good quantum number, matrix elements of both $\mathbf{I}$ and $\mathbf{J}$ must be proportional to matrix elements of $\mathbf{F}$. Thus

$$-g_F \mu_0 \mathbf{F} \cdot \mathbf{H} = -g_J \mu_0 \mathbf{J} \cdot \mathbf{H} - g_I \mu_0 \mathbf{I} \cdot \mathbf{H}. \qquad \text{(VI-7)}$$

We can easily solve for g_F by using the Wigner-Eckart theorem to show that, in an $|l_1 l_2 LM_L\rangle$ scheme, the matrix element of $\mathbf{l}_1$ is equal to the matrix element of $\mathbf{L}$ times $\mathbf{l}_1 \cdot \mathbf{L}/L(L+1)$. Then using the relationship $\mathbf{F} = \mathbf{I} + \mathbf{J}$ to obtain

$$\mathbf{I} \cdot \mathbf{F} = \tfrac{1}{2}(\mathbf{F}^2 + \mathbf{I}^2 - \mathbf{J}^2)$$

and

$$\mathbf{J} \cdot \mathbf{F} = \tfrac{1}{2}(\mathbf{F}^2 + \mathbf{J}^2 - \mathbf{I}^2)$$

we find that

$$g_F = \frac{g_J}{2} \frac{F(F+1) + J(J+1) - I(I+1)}{F(F+1)} + \frac{g_I}{2} \frac{F(F+1) + I(I+1) - J(J+1)}{F(F+1)}.$$

(VI-8)

We note that $g_I \sim (\beta_N/\mu_0)g_J \sim (1/2000)g_J$ so that g_F is effectively given by the first term above. The expectation value for the operator $-g_F \mu_0 \mathbf{F} \cdot \mathbf{H}$ in the low field limit with the external field along the z axis is

$$-g_F \mu_0 M_F H_z.$$

This implies that in the low field limit, each $(2F+1)$ fold degenerate F level will have its degeneracy completely removed.

Eigenvalues of the hyperfine Hamiltonian are also easily expressed in the high field limit. The field dependent terms become simply

$$-g_J \mu_0 M_J H - g_I \mu_0 M_I H,$$

again implying that the former degeneracy is completely removed. Diagonal matrix elements of the remaining terms are obtained by making the replacements.

$$(\mathbf{I} \cdot \mathbf{J}) = M_I M_J,$$

$$3(\mathbf{I} \cdot \mathbf{J})^2 + \tfrac{3}{2}(\mathbf{I} \cdot \mathbf{J}) - I(I+1)J(J+1) = \tfrac{1}{2}(3M_I^2 - I(I+1))(3M_J^2 - J(J+1)),$$

$$10(\mathbf{I} \cdot \mathbf{J})^3 + 20(\mathbf{I} \cdot \mathbf{J})^2$$
$$+ 2(\mathbf{I} \cdot \mathbf{J})[-3I(I+1)J(J+1) + I(I+1) + J(J+1) + 3]$$
$$- 5I(I+1)(J)(J+1)$$
$$= M_I M_J(3I(I+1) - 5M_I^2 - 1)(3J(J+1) - 5M_J^2 - 1).$$

In general, of course, one is in neither the high field nor the low field regime, and a diagonalization of the matrix of the entire hyperfine interaction must be carried out in order to obtain eigenvalues and eigenvectors. This is called the intermediate coupling regime. The classic example[130] of this problem occurs for atoms in the electronic state $^2S_{1/2}$. In this case, selection rules based on J imply that the dipole interaction is the only term in the field-independent part of Eq. (VI-6) which may have non-vanishing matrix elements. Because M ($M = M_I + M_J$ in the high field limit and $M = M_F$ for low fields) is a good quantum number for all fields, the matrix of the hyperfine Hamiltonian (VI-6) must break up into blocks corresponding to basis states of fixed M; each of these blocks can be diagonalized separately. For a fixed M, there can be at most two states, corresponding in low fields to the M_F states arising from $F = I + \frac{1}{2}$ and $F = I - \frac{1}{2}$; therefore, each M block can be at most a 2×2 matrix. For the special case of $M = \pm (I + \frac{1}{2})$, there is only one possible state, this being labeled in low fields as having a total angular momentum $F = I + \frac{1}{2}$. In this case, the M block is only a 1×1 matrix, having the eigenvalue

$$\tfrac{1}{2}AI \mp \tfrac{1}{2}g_J\mu_0 H_z \mp Ig_I\mu_0 H_z, \tag{VI-9a}$$

where the upper signs refer to the case $M = I + \frac{1}{2}$, the lower, to $M = -(I + \frac{1}{2})$. For all other values of M, the Hamiltonian matrix takes the form

$$\begin{vmatrix} \tfrac{1}{2}A(M - \tfrac{1}{2}) - \tfrac{1}{2}g_J\mu_0 H_z - g_I\mu_0(M - \tfrac{1}{2})H_z & \dfrac{A}{2}[I(I+1) - M^2 + \tfrac{1}{4}]^{1/2} \\ \dfrac{A}{2}[I(I+1) - M^2 + \tfrac{1}{4}]^{1/2} & -\tfrac{1}{2}A(M + \tfrac{1}{2}) + \tfrac{1}{2}g_J\mu_0 H_z \\ & \quad - g_I\mu_0(M + \tfrac{1}{2})H_z \end{vmatrix}$$

where the matrix has been constructed using the basis states $|IM_I, \tfrac{1}{2}\tfrac{1}{2}\rangle$ and $|IM_I + 1, \tfrac{1}{2} - \tfrac{1}{2}\rangle$. This matrix has eigenvalues

$$-\dfrac{A}{4} - g_I\mu_0 MH_z \pm \tfrac{1}{2}[A^2(I + \tfrac{1}{2})^2 + (g_J - g_I)^2\mu_0^2 H_z^2$$
$$- 2AM\mu_0 H_z(g_J - g_I)]^{1/2} \tag{VI-9b}$$

which correctly describe the energies of the various states at any value of the magnetic field. The eigenvalues (VI-9) are schematically indicated as a function of magnetic field for $I = 1$ in Figure (VI-1). Such a figure is called a Breit-Rabi diagram. The quantity $h\Delta v$ is defined by the splitting between the $F = I + \frac{1}{2}$ and $F = I - \frac{1}{2}$ states at zero field; using (VI-9) one finds

$$\Delta v = \dfrac{A}{2h}(2I + 1). \tag{VI-10}$$

The Hamiltonian matrix in the general case will be much larger than 2×2 and the eigenvalues cannot be expressed in as simple a form as given above; in general, the matrix must be diagonalized numerically. A typical Breit-Rabi diagram for a somewhat more complicated atom is shown in Figure (VI-2).[131] This diagram illustrates the important point that the F states

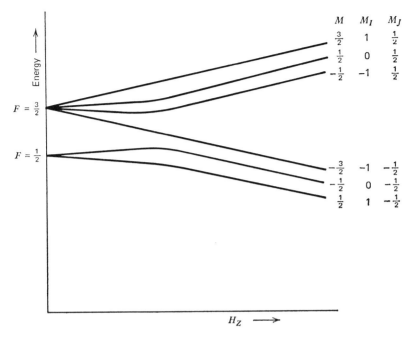

Figure VI-1. Schematic of the energy levels of an atom having $I = 1, J = \tfrac{1}{2}$. The A is assumed to be positive.

are not always ordered nicely at zero field; the exact ordering depends on the values of A, b, and c.

The number of possible **F** values which can be formed by coupling together **I** and **J** is the smaller of $2I + 1$ and $2J + 1$; we call this number $2T + 1$. There are $2T$ independent energy differences between states of different F which can be measured at zero field. There are also $2T$ possible interaction terms in the hyperfine Hamiltonian at zero field (Eq. (IV-18)). The $2T$ interaction constants can therefore be unambiguously determined from the zero field energy differences between states of different F. In practice, of course, one considers only the first two or three terms in the expansion of the hyperfine interactions, making the constants A, b and c overdetermined in many cases. These constants can also be determined equally well in the presence of a weak magnetic field by measuring the differences in energy between states of different F but with $M_F = 0$; in this case the field dependent term in the hyperfine interaction, $g_F \mu_0 M_F H_z$, will vanish. The value of g_I can be easily determined in the high field limit by measuring energy separations between states of constant M_J and different M_I at two known values of the magnetic field; in this case the effect of the field independent hyperfine interactions will be to

128 Hyperfine Structure in an External Field

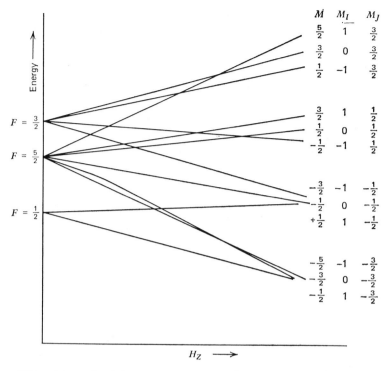

Figure VI-2. Schematic of the energy levels of the $^2P_{3/2}$ state of Ga^{68}. The diagram is drawn assuming that $A \simeq 1.7$ Mc/sec, $b \simeq -10.3$ Mc/sec.

add a constant energy, independent of field, to the separation. This constant will be cancelled out by subtracting the energy difference at one field from the energy difference at the other, so that if $\Delta E(H_1)$ is the separation at field H_1 between states $|IM_I JM_J\rangle$ and $|IM_{I'} JM_J\rangle$, and so on, then

$$\Delta E(H_1) - \Delta E(H_2) = -g_I \mu_0 (M_I - M_{I'})(H_1 - H_2).$$

The constant g_J can be determined in a like manner by measuring energy differences between states of constant M_I and different M_J. Of course, for accurate measurements, one can never assume either weak field or strong field conditions, but rather must always assume intermediate fields and carry out a diagonalization of the entire hyperfine Hamiltonian. This discussion does point out, however, that the values of A, b and c can be determined most accurately from measurements at low fields; those of g_I and g_J, from measurement at high fields.

2. DIAMAGNETIC SHIELDING

In order to obtain the correct value of g_I from measurements of the type discussed in the previous section, one must first correct for the distortion of the atomic core by the external magnetic field. This distortion is very small, and can be discussed from a classical standpoint.

Lamb[132] first considered the effect of the external magnetic field on the atomic core; he made the approximation that an atom is spherically symmetric with charge distribution $\rho(r)$. In this case the amount of charge contained in a volume element taken as a ring with axis passing through the nucleus and parallel to the external field H is

$$\frac{\rho(r)}{2} \sin\theta \, d\theta \, dr.$$

By Larmor's theorem, charges moving in a central field which are suddenly subjected to an external field will simply have a precession of frequency $v_L = eH/4\pi mc$ superimposed on their motion. This precession leads to a current as the charge in the volume element rotates:

$$di = \frac{eH}{4\pi mc} \frac{\rho(r)}{2} \sin\theta \, d\theta \, dr.$$

This induced current produces a magnetic field at the nucleus of magnitude

$$dH'(0) = \frac{2\pi \sin^2\theta}{rc} \frac{eH}{4\pi mc} \frac{\rho(r)}{2} \sin\theta \, d\theta \, dr.$$

Integrating over angle, one finds for the total induced field

$$\frac{H'(0)}{H} = \frac{e}{3mc^2} \int \frac{\rho(r)}{r} dr = \frac{e}{3mc^2} v(0)$$

where $v(0)$ is the potential at the nucleus due to the atomic electrons. This potential is, of course, negative, meaning that the induced field is opposed to the external field; hence, this effect is known as "diamagnetic shielding". The field seen by the nucleus resulting from the application of an external field H is then not H, but rather

$$H\left(1 + \frac{H'(0)}{H}\right).$$

This implies that magnetic moments experimentally determined through interaction with an external magnetic field must be multiplied by a factor

$$\left(1 + \frac{H'(0)}{H}\right)^{-1}$$

to correct for diamagnetic shielding.

Lamb[132] evaluated $v(0)$ using the Thomas-Fermi model and found that

$$\frac{H'(0)}{H} = -0.319 \times 10^{-4} Z^{4/3}.$$

Dickenson[133] evaluated $\sigma = -H'(0)/H$ in the range $1 \leq Z \leq 92$ using Hartree and Hartree-Fock wavefunctions. Kopfermann[134] evaluated

$$\kappa = (1 + H'(0)/H)^{-1}$$

in the same range using Hartree wavefunctions. Bonham and Strand[135] have evaluated σ by employing Hartree-Fock charge densities for small Z and Thomas-Fermi-Dirac densities for large Z. The results using the two different densities converge at approximately $Z = 35$ and apparently are equal at higher Z. For low Z the Hartree-Fock results are smaller in magnitude than those obtained using the statistical potential. The results of Dickenson[134] and Bonham and Strand[135] differ slightly ($<5\%$), with Dickenson having the larger values at large Z, the smaller for the small Z. Malli and Fraga[136] have calculated σ using Hartree-Fock wavefunctions for ground and excited states of neutral atoms and positive and negative ions. They have interpolated their results in order to predict values of σ for all atoms. In addition, they compare values of σ calculated using a variety of methods.

3. EXTERNAL UNIFORM ELECTRIC FIELDS

We shall consider in this section an atom placed in a uniform electric field $\mathbf{E}$; we do not require that the $\mathbf{E}$ field be directed along the z axis. The results of this section can then be combined with the results of the previous section to describe an atom under the influence of electric and magnetic fields arbitrarily oriented with respect to one another; the z axis in such a situation would be defined by the magnetic field. The scalar potential of such an electric field is given by $-\mathbf{E} \cdot \mathbf{r}$; the contribution to the Dirac Hamiltonian is $H_e = e\mathbf{E} \cdot \mathbf{r}$. If we further assume that the electronic wavefunction has a well defined parity and that all of our basis states have the same parity, then expectation values of $e\mathbf{E} \cdot \mathbf{r}$ must vanish because $\mathbf{r}$ has odd parity. (We shall discuss in the next chapter a possible violation of this assumption.) Thus the electric field can only interact with the atomic states in second order (and higher even orders) of perturbation theory. The same conclusion can be reached concerning the interaction of an external electric field with the nucleus; because nuclear states are in general separated by energies large compared to energy shifts obtainable through application of external fields, we can make the approximation that the nucleus does not interact directly with an external electric field.

The second order interaction between the atom and an external electric field is given by

$$H_E = \sum_i \frac{(\psi_0|H_e|\psi_i)(\psi_i|H_e|\psi_0)}{E_0 - E_i}.$$

We wish to study this perturbation in the same approximation as was used in Chapter V; that is, we wish to consider only perturbations for which $E_0 - E_i \approx \Delta E$ for all ψ_0 in the configuration C_0, all ψ_i in the configuration C_i. To further simplify the discussion, we shall assume that the configuration C_0 consists of a number of electrons in filled shells and N electrons in the partially filled nl shell.

The first step in the analysis involves expressing H_e in terms of operators acting on relativistic (nl) states. This is easily done using the techniques developed above:

$$H_e = e \sum \left(\frac{[j, j', S, L]}{[1]}\right)^{1/2} \begin{Bmatrix} \frac{1}{2} & l & j \\ \frac{1}{2} & l' & j' \\ S & L & 1 \end{Bmatrix} (nlj\|rC^{(1)}\|n'l'j')\mathbf{R}^{(SL)1}(nl, n'l') \cdot \mathbf{E}$$

(VI-11)

where the sum is over nlj, $n'l'j'$, and $S + L$ even. Using this expression and the techniques of Chapter V, one finds

$$H_E = e^2 \sum \frac{1}{\Delta E} (nlj_1\|rC^{(1)}\|n'l'j')(n'l'j'\|rC^{(1)}\|nlj_2)$$

$$\times (-1)^{j_1 + j_2 + \phi}[j_1, j_2, S, L]^{1/2} \begin{Bmatrix} \frac{1}{2} & l & j_1 \\ \frac{1}{2} & l & j_2 \\ S & L & k \end{Bmatrix} \begin{Bmatrix} 1 & 1 & k \\ j_1 & j_2 & j' \end{Bmatrix}$$

$$\times \mathbf{R}^{(SL)k}(nl, nl) \cdot (\mathbf{EE})^{(k)}$$

(VI-12)

where the sum is over j_1, j_2, $n'l'j'$, k, and $S + L$ even. The sum over $n'l'$ runs over all closed, as well as empty, shells of the atom; $\phi = 0$ if $n'l'$ is an empty shell, 1 if $n'l'$ is closed. The sum over closed shells is a result of perturbations involving the promotion of an electron from the filled $n'l'$ shell to the nl shell; the sum over open shells results from the promotion of an nl electron to the empty $n'l'$ shell. These interactions are represented by the diagrams

132 Hyperfine Structure in an External Field

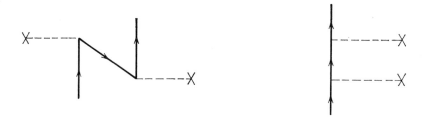

respectively, where X refers to an interaction with an external field. There is also a contribution to H_E of the form

$$\frac{e^2 E^2}{3} \sum \frac{1}{\Delta E} (n_1 l_1 j_1 \| r C^{(1)} \| n_2 l_2 j_2)^2 \qquad \text{(VI-13)}$$

where the sum is over all $n_1 l_1 j_1$ in closed shells, all $n_2 l_2 j_2$ in open (partially filled as well as empty) shells. This contribution results from interactions described by graphs of the type

which illustrate the promotion of an electron from filled to empty shells. (Note: some of the references given below[137-139] do not consider effects described by the graphs

Non-relativistic limits of these equations are easily found. Eq. (VI-12) becomes

$$-\sqrt{2}\, e^2 \sum [l, l'] \begin{pmatrix} l & 1 & l' \\ 0 & 0 & 0 \end{pmatrix}^2 P(nl, n'l')$$

$$\times \begin{Bmatrix} 1 & 1 & k \\ l & l & l' \end{Bmatrix} (-1)^\phi \mathbf{W}^{(0k)k}(nl, nl) \cdot (\mathbf{EE})^{(k)} \qquad \text{(VI-14)}$$

where

$$P(nl, n'l') = \frac{1}{\Delta E} \left| \int R_{nl} R_{n'l'} r \, dr \right|^2$$

and Eq. (VI-13) becomes

$$\frac{2e^2 E^2}{3} \sum P(n_1 l_1, n_2 l_2) \begin{pmatrix} l_1 & 1 & l_2 \\ 0 & 0 & 0 \end{pmatrix}^2 [l_1, l_2]. \quad \text{(VI-15)}$$

In the following discussion, we shall limit our considerations to these nonrelativistic results. Because of the similarity of the relativistic and nonrelativistic forms, however, the discussion can easily be extended by the reader to the relativistic expressions.

The effective interaction H_E given by Eqs. (VI-14) and (VI-15) can be written in the form[137,138]

$$-\tfrac{1}{2}\alpha_{sc} E^2 - \tfrac{1}{2}\alpha_t \mathbf{W}^{(02)2}(nl, nl) \cdot (\mathbf{EE})^{(2)}. \quad \text{(VI-16)}$$

α_{sc}, the scalar polarizability, is defined by

$$\alpha_{sc} = -\frac{e^2}{3} \sum_{n_1 l_1, n_2 l_2} \begin{pmatrix} l_1 & 1 & l_2 \\ 0 & 0 & 0 \end{pmatrix}^2 (2[l_2] - N_{l_2}) N_{l_1} P(n_1 l_1, n_2 l_2)$$

(VI-17)

with N_{l_1} = the number of electrons in the shell $n_1 l_1$, and so on, and α_t, the tensor polarizability, is defined by

$$\alpha_t = 2\sqrt{2}\, e^2 \sum_{n'l'} [l, l'] \begin{pmatrix} l & 1 & l' \\ 0 & 0 & 0 \end{pmatrix}^2 P(nl, n'l')(-1)^\phi \begin{Bmatrix} 1 & 1 & 2 \\ l & l & l' \end{Bmatrix}.$$

(VI-18)

The scalar term $-\tfrac{1}{2}\alpha_{sc} E^2$ shifts all levels of the atom equally, and is therefore of little interest except in third order perturbation theory[137,139,140] (Section VII-6). The tensor term $-\tfrac{1}{2}\alpha_t \mathbf{W}^{(02)2} \cdot (\mathbf{EE})^{(2)}$ can be directly observed because it shifts various levels differently: its expectation value for the level $|\gamma IJFM_F\rangle$ is simply

$$-\frac{\alpha_t}{2}(-1)^{I+J+M_F} \begin{pmatrix} F & 2 & F \\ -M_F & 0 & M_F \end{pmatrix} [F] \begin{Bmatrix} F & 2 & F \\ J & I & J \end{Bmatrix}$$

$$(\gamma J \| W^{(02)2} \| \gamma J)(\mathbf{EE})_0^{(2)}.$$

For electric fields which are small enough to leave J as a good quantum number, it is often convenient to define $\alpha_t(\gamma J)$ as

$$\alpha_t(\gamma J) = (\gamma JJ | W^{(02)2}_0 | \gamma JJ) \sqrt{(\tfrac{4}{6})}\, \alpha_t. \quad \text{(VI-19)}$$

Using this parameter, and evaluating the 3-j and 6-j symbols above, we can express the expectation value of the tensor Stark effect for the level $|\gamma IJFM_F\rangle$ as

$$-\frac{[3M_F^2 - F(F+1)][3X(X-1) - 4F(F+1)J(J+1)]}{(2F+3)(2F+2)(2F)(2F-1)(2J)(2J-1)} \alpha_t(\gamma J)(3E_z^2 - E^2)$$
(VI-20)

where $X = F(F+1) + J(J+1) - I(I+1)$.

Clearly, the external electric field leaves the M_F and $-M_F$ states degenerate. This remaining degeneracy is a result of the electric field (as well as the entire zero-field Hamiltonian) being invariant under time-reversal. Because α_t is a constant for all states in the configuration, $\alpha_t(\gamma J)$'s in different J states should be simply related through Eq. (VI-19):

$$\frac{\alpha_t(\gamma_1 J_1)}{\alpha_t(\gamma_2 J_2)} = \frac{\langle\gamma_1 J_1 J_1|W^{(02)2}_0|\gamma_1 J_1 J_1\rangle}{\langle\gamma_2 J_2 J_2|W^{(02)2}_0|\gamma_2 J_2 J_2\rangle}.$$

Finally, using the results of Chapter IV, one can easily show that the ratio of $\alpha_t(J)$ to the nonrelativistic quadrupole constant b should be constant for all states in the configuration $(nl)^N$:

$$\frac{\alpha_t(\gamma J)}{b} = \left[\frac{5(2l-1)(2l+3)}{6(2l)(2l+1)(l+1)}\right]^{1/2} \frac{\alpha_t}{e^2\langle 1/r^3\rangle Q}.$$

Deviations in these ratios could indicate breakdown of J as a good quantum number, significant relativistic effects in either the hyperfine structure or the Stark effect, or the inadequacy of the treatment of the Stark effect using second-order perturbation theory.

Sums of the type appearing in Eqs. (VI-17) and (VI-18) are often encountered in the study of atomic structure, and it has generally been shown that all states in the sum must be considered in order to get the correct answer. (For a technique to estimate the importance of the higher excited states in the present calculation, see Khadjavi, et al.[138]) Various methods for carrying out these sums have been suggested; an extensive review has been given by Dalgarno.[141] A discussion of these techniques as applied to the calculation of polarizabilities is beyond the scope of this book. However, the techniques discussed in the previous chapter with regards shielding factors and configuration interaction are very similar to some of those used in the polarizability calculations.

Because the expressions given above do not directly involve an interaction with the nucleus, we shall not consider them in any detail. These parameters have been discussed and measured by for example Angel and Sandars,[137] Khadjavi, et al.,[138] Martin, et al.,[142] Feichtner, et al.,[139] and Marrus, et al.[143] We shall, however, use results of this type in Section VII-5 when we discuss third order perturbations involving H_E and H_{hyp}.

4. INTERACTIONS WITH THE GRADIENT OF AN EXTERNAL ELECTRIC FIELD

Both the nucleus and the atomic electrons can interact with the gradient of an electric field. Thus, if the gradient is in the z direction, the interaction will be

$$-\frac{1}{2}\frac{\partial E_z}{\partial z}\left[-e\sum_i C_{i0}^{(2)}r_i^2 + \frac{e}{2}Q\right] \tag{VI-21}$$

where Q is the nuclear quadrupole moment (Eq. (IV-29)). The interaction between the external field and the nucleus follows directly from the results of Chapter IV, and the interaction between the electric field gradient and the electrons follows by analogy.

The interaction between the electrons and the external field is difficult to measure in free atoms because of the quadratic Stark effect. That is, both effects will simultaneously cause energy shifts in the atomic states. The two effects can be separated, however, since the quadrupole interaction and the quadratic Stark shift depend in different ways on the direction of the applied electric field.[144] For example, a reversal of direction of the applied field will effect the quadrupole interaction, which depends on E_z, but not the Stark effect, which depends on E_z^2. In this manner one can measure the atomic quadrupole moment θ, where[144]

$$\theta = -e(JJ|\sum_i r_i^2 C_{i0}^{(2)}|JJ). \tag{VI-22}$$

Typically, the largest field gradient which can be produced in the laboratory is of the order of 5×10^4 volts/cm^2. If we assume an average nuclear quadrupole moment of 5×10^{-24} cm^2, we find a maximum energy shift due to the interaction between the external electric field gradient and the nucleus to be of the order of 10^{-4} cps, which is completely negligible. Much larger gradients exist in nature, however, in for example crystals, in which case the interaction between the crystal electric field gradient and the nucleus may not be negligible.

In any event, if one wishes to study the interaction between the nucleus and an external field gradient, one must consider the distortion of the atomic core by the external gradient. This distortion is essentially the same effect as was considered in Section V-8; in the present case, however, the deformation of the closed shells is caused by the external field gradient rather than by the Coulomb field of the valence shell as was the case in the previous discussion. The calculation of the shielding of the external field gradient can be carried out in the same manner as the calculations described in Section V-8. One defines a shielding constant γ_∞, such that the field gradient at the origin, q, is related to the external field gradient q_e by the equation[145]

$$q = q_e(1 - \gamma_\infty). \tag{VI-23}$$

In terms of a second-order perturbation calculation, γ_∞ can be expressed as

$$\gamma_\infty = -e^2 \sum \frac{1}{\Delta E} \Big\{ (\psi_0 | \sum_i r_i^2 C_{i0}^{(2)} | \psi_j)(\psi_j | \sum_i \frac{C_{i0}^{(2)}}{r_i^3} | \psi_0)$$

$$+ (\psi_0 | \sum_i \frac{C_{i0}^{(2)}}{r_i^3} | \psi_j)(\psi_j | \sum_i r_i^2 C_{i0}^{(2)} | \psi_0) \Big\}.$$

This is the same expression as describes the direct quadrupole shielding term of the last chapter, except that the Coulomb interaction term

$$e^2 \sum_{i<j} \mathbf{C}_i^{(2)} \cdot \mathbf{C}_j^{(2)} r_<^2/r_>^3$$

has been replaced with the term describing the interaction between the field and the atomic electrons (Eq. (VI-21)). Because of the r^2 dependence of the perturbation, the outer levels of the atom are affected most by this induced deformation. Drawing on the results of the quadrupole shielding calculations, we know that this implies that the deformations will be primarily radial, leading to antishielding of the external field. Calculations support this conclusion, with typical values of γ_∞ ranging from -7 to -100.

The calculations of γ_∞ can be carried out using the techniques of Sternheimer described in the previous section. In addition, since the main effect is radial shielding, polarized Hartree-Fock calculations can be expected to give good results. Since, however, this is primarily a problem affecting atoms in solids rather than free atoms, we will not discuss this subject further. The interested reader can find discussions of these calculations and tables of calculated values of γ_∞ given by for example Sternheimer and coworkers,[145] Freeman and Watson,[113,146] Lahiri et al.,[147] Dalgarno,[141,148] Langhoff and Hurst,[149] Kelly and Taylor,[150] and Gupta, et al.[150a]

VII. Higher-Order Effects

1. INTRODUCTION

In this chapter we consider a variety of effects loosely lumped together under the name "higher-order effects." These effects as a whole are really higher-order only in the sense that they are smaller than the effects discussed in the previous chapters. Some of them, like the Bohr-Weisskopf effect (Section 2) are parts of the first-order hyperfine interaction which we have previously neglected. Others arise from the second-order hyperfine effects (Section VII-3). Although in general small, the terms discussed in this chapter can influence greatly the interpretation made of experimental data. In addition, some of these effects, although reasonably simple in origin, can be the most difficult part of the interaction to evaluate with any accuracy, and may provide the most exacting test of the models used.

2. HYPERFINE ANOMALIES

The dipole interaction constant A was defined in Section IV-5 by

$$A = \frac{\mu_I}{IJ}(JJ|M_{D0}^{(1)}|JJ) \tag{IV-24}$$

where $(JJ|M_{D0}^{(1)}|JJ)$ becomes, in the non-relativistic limit, the negative of the magnetic field at the nucleus produced by the orbital electrons. According to the central-field approximation of Chapter III, the electronic wavefunctions depend on only one property of the nucleus: its total charge Z. Thus two isotopes of the same element should have the same electronic wavefunctions and the matrix element of $(JJ|M_{D0}^{(1)}|JJ)$ should be the same for both isotopes. In such a case, the relationships

$$\frac{A(1)}{A(2)} = \frac{\mu_I(1)I(2)}{\mu_I(2)I(1)} = \frac{g_I(1)}{g_I(2)} \tag{VII-1}$$

should hold, where $A(1)$, $\mu_I(1)$, and so on, are the dipole interaction constant and dipole moment of isotope 1, and so on. In the previous chapter, it was shown that all the quantities appearing in this equation can be measured; however, when these measured values are used, Eq. (VII-1) is often found to be incorrect. Deviations from the simple relationship suggested by Eq. (VII-1) are expressed in terms of a hyperfine anomaly Δ_{12}:

$$\Delta_{12} = \frac{A(1)g_I(2)}{A(2)g_I(1)} - 1. \qquad \text{(VII-2)}$$

The anomaly, although generally of the order of a fraction of a percent, can be almost as large as 10%. Recent compilations of measured anomalies have been given by Stroke, et al.[151,152], Foley,[153] and Fuller and Cohen.[154]

Thus far in our discussion of fine and hyperfine structure we have twice assumed that the nucleus is a point charge, and the error involved in both of these assumptions can contribute to Δ_{12}. The assumption was first made in the discussion of fine structure where the potential of the nucleus was assumed to be Ze/r all the way to the origin. In reality, the nucleus has a finite volume and the potential deviates very much from Coulombic inside this volume. Furthermore, this volume may be different for different isotopes of the same element, so that the electronic wavefunctions of two isotopes will not necessarily be equal. Errors arising from the assumption of a completely Coulombic potential were first discussed by Rosenthal and Breit.[155] Our second assumption concerning the size of the nucleus was made in determining the vector potential of the nucleus (Eq. (IV-11)). Here, we made the assumption that the electron was always outside the nucleus ($r_e > r_n$); the definition (IV-24) of A was based on this assumption. Clearly, the use of the total vector potential would change the definition of A. The effect of using the total vector potential was first considered by Bohr and Weisskopf.[156]

There are two electronic reduced-mass corrections to the magnetic interaction constants. The first is the isotopic shift reduced-mass correction to the electronic wavefunction which effectively reduces $(1/r^k)$-like terms by a factor $(1 + m/AM)^{-k}$, where m = electron mass, M = proton mass. The second is a reduction in the mass appearing in the electronic orbital g factor. There is also a reduced mass correction to the nucleon orbital g factor. All of these reduced mass corrections are important for the lighter elements only and we shall consider them no further.

a. The Rosenthal-Breit Effect[155]

The potential felt by an electron at moderate and large r is independent of whether the nucleus is a point or has a finite volume. Thus we would expect wavefunctions obtained using a potential appropriate to an extended nucleus

(F_{nlj}^e and G_{nlj}^e) to differ only at small values of r from the wavefunctions obtained for a point nucleus. Let us now consider a one-electron matrix element of the kth term in the magnetic hyperfine interaction, evaluated using extended nucleus wavefunctions. Such a matrix element is simply related to the usual matrix element evaluated with point nucleus wavefunctions by the equation

$$(I(\tfrac{1}{2}l)j|\mathbf{M}_D^{(k)} \cdot \mathbf{N}^{(k)}|I(\tfrac{1}{2}l)j)_{\text{extended}} = (I(\tfrac{1}{2}l)j|\mathbf{M}_D^{(k)} \cdot \mathbf{N}^{(k)}|I(\tfrac{1}{2}l)j)_{\text{point}} \times (1 - \phi_{lj}{}^k)$$

(VII-3)

where

$$\phi_{lj}{}^k = 1 - \left(\int \frac{F_{nlj}^e G_{nlj}^e}{r^{k+1}} \, dr \bigg/ \int \frac{F_{nlj} G_{nlj}}{r^{k+1}} \, dr \right).$$

Obviously, $\phi_{lj}{}^k$ will be negligibly small unless the largest contribution to the integrals comes from small r where F_{nlj}^e and F_{nlj}, etc., differ the most. A brief discussion of methods commonly used to calculate $\phi_{lj}{}^k$ should make it obvious that this condition will only be satisfied for certain combinations of k, l and j.

We assume for the moment that all electrons except the one of interest are in closed, and therefore spherical, shells. The Rosenthal-Breit effect is felt by an electron only when the electron is inside, or very near to, the nucleus; at such time the electron can be considered to be inside of the orbits of all of the remaining electrons of the atom. By the laws of elementary electrostatics, under such circumstances the electron of interest moves in a field produced entirely by the nucleus. Inside the nucleus, the potential can be expressed in the form $\sum_b a_b r^b$; immediately outside the nucleus the field is, of course, Ze/r. If the major part of the integral $\int (F_{nlj}^e G_{nlj}^e / r^{k+1}) \, dr$ comes from small r, then the integral can be evaluated reasonably well by using wavefunctions calculated for an electron moving in these simple fields. Of course, wavefunctions obtained in this way will be inaccurate at larger r where the effects of shielding of the nuclear charge by the closed shells would be felt, so that the contribution to the integral from large r will be in error. However, if we calculate the point nucleus wavefunctions by putting an electron in a Coulombic field of Ze/r, the contribution to $\int (F_{nlj} G_{nlj}/r^{k+1}) \, dr$ from large r will be approximately the same as the contribution to $\int (F_{nlj}^e G_{nlj}^e/r^{k+1}) \, dr$ from large r. If this contribution is small compared to the contributions to the integrals from small r, it will approximately drop out of a ratio of the two integrals such as appears in Eq. (VII-3). We should be able to calculate $\phi_{lj}{}^k$ much more accurately using these techniques, therefore, than we could calculate either integral separately.

In Chapter VIII we discuss solutions of the point nuclear equation in the approximation of zero binding energy ($E - mc^2 \approx 0$). We consider here the solution to the extended nuclear problem in the same approximation. In this approximation, the radial equations (III-11) can be written

$$\left(\frac{d}{dr} - \frac{\kappa}{r}\right)F = -\frac{1}{\hbar c}(2mc^2 + eV)G$$

$$\left(\frac{d}{dr} + \frac{\kappa}{r}\right)G = -\frac{1}{\hbar c}(-eV)F.$$

The potential inside the nucleus can be expanded in a series $\sum_b a_b r^b$ where the constants a_b depend on the nuclear model chosen. For the sake of this example we will assume a uniformly charged sphere of radius R_N; the appropriate potential is

$$V = \left[\frac{3}{2} - \frac{1}{2}\left(\frac{r}{R_N}\right)^2\right]\frac{Ze}{R_N}.$$

The functions F and G can be expanded for $r \leq R_N$ in a series

$$F^e = \sum_n c_n \left(\frac{r}{R_N}\right)^n$$

$$G^e = \sum_n d_n \left(\frac{r}{R_N}\right)^n \qquad \text{(VII-4a)}$$

where $n \geq 0$ to assure proper behavior at the origin. Inserting this expansion and the above potential into the radial equations and picking out the coefficients of $(r/R_N)^{n-1}$ leads to the conditions on the coefficients

$$c_n(n - \kappa) = -\frac{1}{\hbar c}\left[\left(2mc^2 R_N + \frac{3}{2}Ze^2\right)d_{n-1} - \frac{Ze^2}{2}d_{n-3}\right]$$

$$d_n(n + \kappa) = -\frac{1}{\hbar c}\left[-\frac{3}{2}Ze^2 c_{n-1} + \frac{Ze^2}{2}c_{n-3}\right]. \qquad \text{(VII-4b)}$$

Outside the nucleus, these solutions are matched to the solutions for the point nucleus given in Section VIII-2. Normalization conditions set the absolute value of the c and d coefficients.

Using the results given above, one can determine the conditions under which the extended nucleus integrals may have large contributions from the area near the origin. For the smallest value of n which is allowed, n_0, we have from Eq. (VII-4b):

$$c_{n_0}(n_0 - \kappa) = 0$$
$$d_{n_0}(n_0 + \kappa) = 0.$$

Solutions to these equations are: if $j = l + \frac{1}{2}$, then $n_0 = l + 1$ and $d_{n0} = 0$; if $j = l - \frac{1}{2}$, then $n_0 = l$ and $c_{n0} = 0$. Thus the integrals taken over the nuclear volume have the form

$$\int_0^{R_N} \frac{F_{nlj}^e G_{nlj}^e}{r^{k+1}} dr \sim \int_0^{R_N} \frac{r^{2|\kappa|+1}}{r^{k+1}} dr = \int_0^{R_N} \frac{r^{2j+1}}{r^k} dr.$$

For $R_N/a_0 \ll 1$, this integral will be largest when $k = k_{\max} = 2j$. Using the results of Section VIII-2, one can also show that the same conditions indicate that the radial integrals which appear in the usual magnetic interactions also may have large contributions from the area near to the origin. We find, then, that the radial integrals of interest will most likely have large contributions near the origin if $k = 2j$; following the arguments given above, only the values ϕ_{lj}^{2j} will be significantly different from zero. When $k = 1$, this effect is known as the Breit-Rosenthal effect, and is important for $s_{1/2}$ and $p_{1/2}$ electrons. We shall return to a discussion of this effect below, after considering other mechanisms which may affect the dipole interaction constant.

b. The Bohr-Weisskopf Effect[156]

We now consider the effect of neglecting the part of the vector potential for $r_e < r_n$. The correct form of the magnetic contribution to H_{hyp} can be written

$$\sum_k H^k = ie\beta_N \sum_{k,n} \left[\sqrt{\left(\frac{2k-1}{k+1}\right)} \frac{r_n^{k-1}}{r_e^{k+1}} (\boldsymbol{\alpha}_e \mathbf{C}_e^{(k)})^{(k)} \cdot (2g_{ln}(\mathbf{C}_n^{(k-1)}\mathbf{L}_n)^{(k)} \right.$$

$$\left. + (k+1)g_{sn}(\mathbf{C}_n^{(k-1)}\mathbf{S}_n)^{(k)}) \right]$$

$$+ ie\beta_N \sum_{k,n} \left[\sqrt{\left(\frac{2k+3}{k}\right)} \frac{r_e^k}{r_n^{k+2}} (\boldsymbol{\alpha}_e \mathbf{C}_e^{(k)})^{(k)} \cdot (2g_{ln}(\mathbf{C}_n^{(k+1)}\mathbf{L}_n)^{(k)} \right.$$

$$\left. - kg_{sn}(\mathbf{C}_n^{(k+1)}\mathbf{S}_n)^{(k)}) \right] \quad \text{(VII-5)}$$

where the first term applies if $r_e > r_n$, the second if $r_n > r_e$. The magnetic hyperfine interaction due to a point nucleus, $\sum_k \mathbf{M}^{(k)} \cdot \mathbf{N}^{(k)}$, is, of course, given by the first term above, and it was this expression that was used in the discussions in the latter part (Sections 2–6) of Chapter IV. Since $\mathbf{M}^{(k)} \cdot \mathbf{N}^{(k)}$ is a special case of H^k, a matrix element of H^k between one electron states can be related to the corresponding matrix element of $\mathbf{M}^{(k)} \cdot \mathbf{N}^{(k)}$. Bohr and Weisskopf[156] considered the relationship between matrix elements of these operators using extended nucleus wavefunctions:

$$(I(\tfrac{1}{2}l)j|H^k|I(\tfrac{1}{2}l)j)_{\text{extended}} = (I(\tfrac{1}{2}l)j|\mathbf{M}^{(k)} \cdot \mathbf{N}^{(k)}|I(\tfrac{1}{2}l)j)_{\text{extended}}(1 - \varepsilon_{lj}^k)$$

where

$$\varepsilon_{lj}^K = \beta_N \Bigg[\bigg(\frac{2}{K+1}\sqrt{(2K-1)K}\sum_n \int_{\tau_N} \psi_I^* \psi_I \, d\tau_N \, g_{ln}(\mathbf{C}_n^{(K-1)}\mathbf{L}_n)_0^{(K)} \int_0^{r_n} F_{nlj}^e G_{nlj}^e$$

$$\times \bigg(\frac{r_n^{K-1}}{r_e^{K+1}} - \frac{r_e^K}{r_n^{K+2}}\bigg) dr_e\bigg)\bigg(N_I^K \int_0^\infty \frac{F_{nlj}^e G_{nlj}^e}{r_e^{K+1}} dr_e\bigg)^{-1}\Bigg]$$

$$+ \beta_N \Bigg[\bigg(\sum_n \int_{\tau_N} \psi_I^* \psi_I \, d\tau_N \bigg(\sqrt{(2K-1)K} \, g_{sn}(\mathbf{C}_n^{(K-1)}\mathbf{S}_n)_0^{(K)} \int_0^{r_n} F_{nlj}^e G_{nlj}^e$$

$$\times \frac{r_n^{K-1}}{r_e^{K+1}} dr_e + \sqrt{\bigg(\frac{(2K+3)K^2}{K+1}\bigg)} g_{sn}(\mathbf{C}_n^{(K+1)}\mathbf{S}_n)_0^{(K)} \int_0^{r_n} F_{nlj}^e G_{nlj}^e$$

$$\times \frac{r_e^K}{r_n^{K+2}} dr_e\bigg)\bigg)\bigg(N_I^K \int_0^\infty \frac{F_{nlj}^e G_{nlj}^e}{r_e^{K+1}} dr_e\bigg)^{-1}\Bigg]$$

(VII-6)

with $N_I^K = (II|N_0^{(K)}|II)$. The first term in square brackets above is the change in the matrix element produced by the orbital motion of the nucleons; the second, the change produced by the nucleon spin. We have used the relationship

$$(C^{(K+1)}L)^{(K)} = \bigg[\frac{K(2K-1)}{(K+1)(2K+3)}\bigg]^{1/2}(C^{(K-1)}L)^{(K)}$$

to simplify the results for the orbital interaction. Using the equations obtained above for F_{nlj}^e and G_{nlj}^e, one can easily show that the only terms of importance in ε_{lj}^k will be those with $k = 2j$. The term with $k = 1$ gives rise to the Bohr-Weisskopf effect. We shall consider this effect in greater detail below.

c. Application to the Many-Electron Problem

We find, then, that for our purposes the only necessary corrections to diagonal one-electron matrix elements of the magnetic hyperfine interaction are given by Eqs. (VII-3) and (VII-6). Corrections to the many-electron Hamiltonian are easily made: the factor $(1 - \phi_{ij}^k)(1 - \varepsilon_{ij}^k)$ should be inserted inside the sum over j and k in the part of Eq. (IV-19b) diagonal in l and j. In principle, there is also a correction to matrix elements off-diagonal in l and j, but since these are small effects, we can neglect off-diagonal parts. Thus, we can correct the results of Eq. (IV-20) for the effects discussed in this section by replacing P_{++} by $P_{++}(1-\delta(l,0)\phi_{0\frac{1}{2}}^1)(1-\delta(l,0)\varepsilon_{0\frac{1}{2}}^1)$ and P_{--} by $P_{--}(1-\delta(l,1)\phi_{1\frac{1}{2}}^1)$ $\times (1 - \delta(l,1)\varepsilon_{1\frac{1}{2}}^1)$. The results of Eq. (IV-22) can be corrected by replacing U_{++} by $U_{++}(1 - \delta(l,1)\phi_{1\frac{1}{2}}^3)(1 - \delta(l,1)\varepsilon_{1\frac{1}{2}}^3)$ and U_{--} by $U_{--}(1 - \delta$

$(l, 2)\phi_{2\frac{2}{3}}{}^{3})(1 - \delta(l, 2)\varepsilon_{2\frac{3}{2}}{}^{3})$. In light of the discussions given above and in Chapter IX, one would expect the factors $\phi_{j+\frac{1}{2}j}{}^{2j}$ and $\varepsilon_{j+\frac{1}{2}j}{}^{2j}$ to be roughly $(Z\alpha)^2$ as large as the factors $\phi_{j-\frac{1}{2}j}{}^{2j}$ and $\varepsilon_{j-\frac{1}{2}j}{}^{2j}$. Thus, as Z increases, the corrections to P_{--} and U_{--} will increase in importance relative to the corrections to P_{++} and U_{++}, although the former corrections will always remain smaller than the latter.

The operators $\mathbf{M}^{(1)}$ and $\mathbf{M}^{(3)}$ modified as described above can be used to calculate the interaction constants A and c for an extended nucleus. We have mentioned before the difficulty in obtaining accurate radial electronic wavefunctions. In order to calculate a meaningful change in A or c in going from an interaction with a point nucleus to one with an extended nucleus, the wavefunctions must be accurate to better than a fraction of a percent, the order of magnitude of a typical ε^1. Except in a few cases, primarily very light atoms, such accuracy is impossible. On the other hand, it is not too unlikely than in many instances ratios of A's (or c's) can be calculated to such an accuracy. Thus we can hope to study theoretically anomalies as defined by Eq. (VII-2) for the dipole interaction. One can also define an anomaly for the octupole interaction

$$\frac{c(1)\Omega(2)}{c(2)\Omega(1)} - 1 = \delta^{12}.$$

Of course, such an equation may be of very limited use since the octupole moments are exceedingly difficult to measure directly because they interact only with the second spatial derivative of an external magnetic field and they appear to be very small (10^{-25} cm^2). We shall therefore restrict the remainder of our discussion to dipole anomalies.

It is clear that relating measured values of Δ_{12} to calculated values of $\phi_{lj}{}^\kappa$ and $\varepsilon_{lj}{}^\kappa$ is not easy; qualitatively the relationship is of the form

$$\frac{A(1)}{A(2)} = \frac{g_I(1)[M(l) + M(s)(1 - \phi_0(1))(1 - \varepsilon_0(1)) + M(p)(1 - \alpha\phi_1(1))(1 - \alpha\varepsilon_1(1))]}{g_I(2)[M(l) + M(s)(1 - \phi_0(2))(1 - \varepsilon_0(2)) + M(p)(1 - \alpha\phi_1(2))(1 - \alpha\varepsilon_1(2))]}$$

$$= \frac{g_I(1)}{g_I(2)}(1 + \Delta_{12})$$

where $M(l)$ is the part of $(JJ|M_{D0}^{(1)}|JJ)$ obtained by restricting the sum over l in $M_{D0}^{(1)}$ to $l > 1$; $M(s)$, that part of the sum with $l = 0$; and $M(p)$, that part of the sum with $l = 1$. We have written the value of $\varepsilon_{1\frac{1}{2}}{}^1$ for isotope 1 as $\varepsilon_l(1)$, and so on, for conciseness. The factor α is related to the probability of finding a p electron in the $j = \frac{1}{2}$ state. The situation is further complicated, however, by core polarization effects. As is shown in a previous chapter, a considerable s electron density can be produced by second-order perturbation interactions. This second-order s electron density can also contribute to Δ_{12}. In the case where the atom of interest has no s or p electrons in partially filled shells, any measured anomaly can be attributed to core polarization effects.

d. Discussion

A number of comprehensive calculations of ϕ_l and ε_l have been carried out utilizing a variety of nuclear models. Rosenthal and Breit[155] based their calculation of ϕ_l on a nuclear model in which the nuclear charge was distributed uniformly on the surface of the nucleus. Crawford and Schawlow[157] considered a nucleus with the charge distributed uniformly through the volume of the nucleus. The results of both works can be expressed as

$$\phi_l = \frac{2(\kappa+\rho)(\rho)(2\rho+1)}{(2\kappa+1)[\Gamma(2\rho+1)]^2}\left(\frac{2pZR_N}{a_0}\right)^{2\rho-1}$$

where p is a constant of order 1, and $\rho = \sqrt{\kappa^2 - (Z\alpha)^2}$. Calculations of ε_l are carried out by using the expansion for F^e and G^e given by Eq. (VII-4a); then

$$\varepsilon_l = \frac{\beta_N}{\mu_I}\sum_n \int_{\tau_N} \psi_I^* \psi_I \sum_m \left(\frac{r_n}{R_N}\right)^{2m}$$
$$\times \left[g_{sn}S_{nz}(b_S^l)_{2m} + \frac{\sqrt{(10)}}{2}(\mathbf{S}_n\,\mathbf{C}_n^{(2)})_0^{(1)}(b_D^l)_{2m} + g_{ln}L_{nz}(b_L^l)_{2m}\right]d\tau_N$$

where b^l's are functions of the c and d coefficients of Eq. (VII-4), R_N, and $(\int_0^\infty F_{nlj}^e G_{nlj}^e / r^2\,dr)^{-1}$. Bohr and Weisskopf[156] calculated the b^l's by assuming a nucleus with uniform charge distribution. They also assumed a spherical distribution of nuclear spin magnetic moment, allowing them to drop the term containing $(\mathbf{SC}^{(2)})^{(1)}$. Values of $b(s)$ and $b(p)$, quantities related to the b^l's by[158]

$$\sum_m \left(\frac{r_n}{R_N}\right)^{2m}(b_S^l)_{2m} = b(l)\left(\frac{r_n}{R_N}\right)^2$$
$$\sum_m \left(\frac{r_n}{R_N}\right)^{2m}(b_L^l)_{2m} = 0.62b(l)\left(\frac{r_n}{R_N}\right)^2$$

are tabulated by Bohr and Weisskopf for $Z = 10$–90 in steps of 10. These values were used by Reiner[159] in conjunction with the collective model to predict values of ε_l. Eisenger and Jaccarino[160] use the same charge distribution but make no assumption as to the distribution of nuclear magnetism. They have tabulated b^l coefficients for $Z = 10$–90 in steps of 5 and have discussed evaluation of nuclear integrals using the single particle model of the nucleus. Stroke, Blin-Stoyle and Jaccarino[151] have calculated values of $(1 - \phi_l)\varepsilon_l$ (a parameter obtained by changing the integrals in the denominator of the right hand side of Eq. (VII-6) to $\int F_{nlj}G_{nlj}/r^{K+1}\,dr$). Using a potential based on electron scattering results, they have given an extensive tabulation of b^l coefficients as a function of A from $Z = 17$–100 in steps of various sizes. Their discussion of evaluation of the nuclear integrals is centered on the configuration mixing model.

The results obtained by Rosenthal and Breit,[155] Crawford and Shawlow[157] show that, although ϕ_l can be as large as 10–20%, $\phi_l(1)$ and $\phi_l(2)$ will differ only through differences in $R_N(1)$ and $R_N(2)$. Thus, the contribution to Δ_{12} from ϕ_l can be expected to be small ($\sim \phi_l(1) - \phi_l(2) \sim 10^{-4}$). The calculated values of ε_l show that this quantity, whose magnitude is very roughly given by the equation[156]

$$\varepsilon_l \sim -\left(\frac{ZR_n}{a_0}\right)\left(\frac{a_0}{2ZR_N}\right)^{2(1-\rho)}\left(\frac{r_n^{\,2}}{R_N^{\,2}}\right)_{\text{avg}}(Z\alpha)^{2l},$$

is somewhat smaller (~ 5–10% for very large Z) than ϕ_l. This quantity depends much more strongly on the properties of the nucleus than does ϕ_l, however, because the value of the nuclear operator depends not only on the radius of the nucleus, but also on the spin and orbital states of the nucleons. Thus if the nucleons in two isotopes are in quite different orbital and spin states, the matrix elements of S_{nz}, L_{nz} and $(\mathbf{S}_n \mathbf{C}_n^{(2)})_0^{(1)}$ may be quite different. This means that $\varepsilon_l(1)$ and $\varepsilon_l(2)$ may be significantly different even though the radial integrals used to calculate these quantities are approximately the same. All of this indicates that if a large anomaly is observed, it is probably due to the Bohr-Weisskopf effect, but if the anomaly is small, it could be due to either the Bohr-Weisskopf or Rosenthal-Breit effects. Finally, let us note that these results are independent of two parameters, the principal quantum number n and the degree of ionization of the atom, which at first thought might seem to be of some importance. The former parameter disappeared from the calculation because we assumed the electron to have zero binding energy. This assumption should be reasonably valid for large n (Kopfermann[134] suggests "large" to be $n \geq 6$), but might possibly cause these calculations to be in error for small n. The degree of ionization disappeared because we assumed that the principal contributions to the integrals of interest arose from regions inside of the orbits of all the remaining electrons. Fortunately, experimental evidence seems to attest to the validity of these results. For example, Bordarier, et al.[77] conclude that the anomaly produced by s electrons is constant for EuI $4f^7 6s^2$, EuIII $4f^7$, and several levels of EuI $4f^7 6s6p$.

3. BREAKDOWN OF J AS A GOOD QUANTUM NUMBER

In previous work, we have assumed that J is a good quantum number. In this section, we shall investigate the results of removing this requirement in second order of perturbation theory. We will consider the breakdown of J

due to the hyperfine interaction, H_{hyp}, (Eq. (IV-18)) and the interaction of the electrons with an external magnetic field (Eq. (VI-2)), which we shall abbreviate as $\mathbf{P}^{(1)} \cdot \mathbf{H}$. The second order terms of interest are given by

$$\sum_{\psi_i} \frac{(\psi_0|H_{\text{hyp}}|\psi_i)(\psi_i|H_{\text{hyp}}|\psi_0')}{E_0 - E_i}, \qquad \text{(VII-7a)}$$

$$\sum_{\psi_i} [(\psi_0|H_{\text{hyp}}|\psi_i)(\psi_i|\mathbf{P}^{(1)} \cdot \mathbf{H}|\psi_0') + (\psi_0|\mathbf{P}^{(1)} \cdot \mathbf{H}|\psi_i)(\psi_i|H_{\text{hyp}}|\psi_0')] \frac{1}{E_0 - E_i}, \qquad \text{(VII-7b)}$$

and

$$\sum_{\psi_i} \frac{(\psi_0|\mathbf{P}^{(1)} \cdot \mathbf{H}|\psi_i)(\psi_i|\mathbf{P}^{(1)} \cdot \mathbf{H}|\psi_0')}{E_0 - E_i} \qquad \text{(VII-7c)}$$

where the states ψ_i are states of the same configuration as the unperturbed state ψ_0, but $\psi_i \neq \psi_0$. The study can be carried out in either the low- or high-field limits; if the hyperfine energies are assumed small compared to the fine structure, the result must be independent of the coupling between $\mathbf{I}$ and $\mathbf{J}$. Parts a and b of Eqs. (VII-7) will be considered in the low field limit $(I + J = F)$, part c in the high field limit. Woodgate[100] has carried out equivalent derivations in the high field limit for all three cases.

For convenience, we write

$$H_{\text{hyp}} = \sum \mathbf{A}_n^{(k)} \cdot \mathbf{B}_e^{(k)}$$

where $\mathbf{A}_n^{(k)}$ operates in the nuclear space, $\mathbf{B}_e^{(k)}$ in the electron space. Of course, for k even, $\mathbf{A}_n^{(k)} \cdot \mathbf{B}_e^{(k)}$ is equal to $\mathbf{F}^{(k)} \cdot \mathbf{Q}^{(k)}$; for k odd, $\mathbf{N}^{(k)} \cdot \mathbf{M}^{(k)}$. The interaction given by Eq. (VII-7a) can then be written

$$\sum_{\substack{J' \\ K, k}} (IJFM|\mathbf{A}_n^{(K)} \cdot \mathbf{B}_e^{(K)}|IJ'FM)(IJ'FM|\mathbf{A}_n^{(k)} \cdot \mathbf{B}_e^{(k)}|IJFM) \frac{1}{E_J - E_J'}$$

$$= (-1)^{2I+J+J'+2F} \begin{Bmatrix} I & J' & F \\ J & I & K \end{Bmatrix} \begin{Bmatrix} I & J & F \\ J' & I & k \end{Bmatrix} (I\|A_n^{(K)}\|I)$$

$$\times (I\|A_n^{(k)}\|I)(J\|B_e^{(K)}\|J')(J'\|B_e^{(k)}\|J) \frac{1}{E_J - E_J'}.$$

We can use the Biedenharn-Elliott sum rule (Eq. (II-12)) to rewrite the above in the form

$$\sum_\kappa (-1)^{I+J+F} \begin{Bmatrix} J & I & F \\ I & J & \kappa \end{Bmatrix} (I\|G_n^{(\kappa)}\|I)(J\|H_e^{(\kappa)}\|J)$$

where

$$(I\|G_n^{(\kappa)}\|I)(J\|H_e^{(\kappa)}\|J) = \sum_{kKJ'} (-1)^{2I+2J+k+K+\kappa} \frac{[\kappa]}{E_J - E_J'} \begin{Bmatrix} K & k & \kappa \\ I & I & I \end{Bmatrix} \begin{Bmatrix} K & k & \kappa \\ J & J & J' \end{Bmatrix}$$

$$(I\|A_n^{(K)}\|I)(I\|A_n^{(k)}\|I)(J\|B_e^{(K)}\|J')(J'\|B_e^{(k)}\|J). \qquad \text{(VII-8)}$$

This last form of the interaction is simply equal to

$$(IJFM|\sum \mathbf{G}_n^{(\kappa)} \cdot \mathbf{H}_e^{(\kappa)}|IJFM)$$

implying that, to second order, the hyperfine interaction can be written as

$$H'_{\text{hyp}} = \sum_k (\mathbf{F}^{(k)} \cdot \mathbf{Q}_D^{(k)} + \mathbf{N}^{(k)} \cdot \mathbf{M}_D^{(k)} + \mathbf{G}_n^{(k)} \cdot \mathbf{H}_e^{(k)}).$$

The hyperfine constants which will be measured will then differ from the constants of Section IV-5. In particular, the measured dipole constant A_m is related to the "real" A of Section IV-5 by

$$A_m = A + \frac{1}{IJ}(II|G_{n0}^{(1)}|II)(JJ|H_{e0}^{(1)}|JJ); \qquad \text{(VII-9a)}$$

the measured quadrupole constant b_m to b by

$$b_m = b + 4(II|G_{n0}^{(2)}|II)(JJ|H_{e0}^{(2)}|JJ); \qquad \text{(VII-9b)}$$

and the measured octupole constant c_m to c by

$$c_m = c + (II|G_{n0}^{(3)}|II)(JJ|H_{e0}^{(3)}|JJ). \qquad \text{(VII-9c)}$$

The perturbation of Eq. (VII-7b) will be linear in the external field, and so may effect measured values of g_J and g_I. The perturbation is given by

$$\sum [(IJFM|\mathbf{P}^{(1)} \cdot \mathbf{H}|IJ'F'M')(IJ'F'M'|\mathbf{A}_n^{(k)} \cdot \mathbf{B}_e^{(k)}|IJF'M')$$

$$+ (IJFM|\mathbf{A}_n^{(k)} \cdot \mathbf{B}_e^{(k)}|IJ'FM)(IJ'FM|\mathbf{P}^{(1)} \cdot \mathbf{H}|IJF'M')]\frac{1}{E_J - E'_J}$$

$$= \sum_{\substack{qk \\ J'}} (-1)^{F-M+q} \begin{pmatrix} F & 1 & F' \\ -m & q & m' \end{pmatrix} H_{-q} (J\|P^{(1)}\|J')(J'\|B_e^{(k)}\|J)(I\|A_n^{(k)}\|I)$$

$$\times \left[(-1)^{2I+2J'+F+F'+1}[F,F']^{1/2} \begin{Bmatrix} F & 1 & F' \\ J' & I & J \end{Bmatrix} \begin{Bmatrix} I & J & F' \\ J' & I & k \end{Bmatrix} \right.$$

$$\left. + (-1)^{2I+2J+2F+1}[F,F']^{1/2} \begin{Bmatrix} I & J' & F \\ J & I & k \end{Bmatrix} \begin{Bmatrix} F & 1 & F' \\ J & I & J' \end{Bmatrix} \right] \frac{1}{E_J - E'_J}$$

$$= \sum_{\substack{q\kappa k \\ J'}} (-1)^{F-M+q} \begin{pmatrix} F & 1 & F' \\ -M & q & M' \end{pmatrix} H_{-q}(J\|P^{(1)}\|J')(I\|A_n^{(k)}\|I)$$

$$\times (J'\|B_e^{(k)}\|J)[F,F']^{1/2}[\kappa] \begin{Bmatrix} I & I & k \\ J & J & \kappa \\ F & F' & 1 \end{Bmatrix} \begin{Bmatrix} J & J & \kappa \\ 1 & k & J' \end{Bmatrix}$$

$$\times \frac{(-1)^{2J+1}}{E_J - E'_J}(1 + (-1)^{1+\kappa+k})$$

$$= \sum_{\kappa k} (IJFM|(\mathbf{A}_n^{(k)}\mathbf{T}_e^{(\kappa)})^{(1)} \cdot \mathbf{H}|IJF'M')$$

where

$$(IJ\|A_n^{(k)}T_e^{(\kappa)}\|IJ) = \sum_{J'} (J\|P^{(1)}\|J')(J'\|B_e^{(k)}\|J)(I\|A_n^{(k)}\|I)$$

$$\times \frac{[\kappa]}{\sqrt{3}} \frac{(-1)^{2J+1}}{E_J - E_J'} \begin{Bmatrix} 1 & k & \kappa \\ J & J & J' \end{Bmatrix} (1 + (-1)^{1+\kappa+k}). \quad \text{(VII-10)}$$

We have again used the Biedenharn-Elliott sum rule in going from the second to the third step above. We conclude, then, that an operator

$$\sum_{k,\kappa} (\mathbf{A}_n^{(k)}\mathbf{T}_e^{(\kappa)})^{(1)} \cdot \mathbf{H}$$

must be added to the Hamiltonian to account for second order interactions between the Zeeman and hyperfine effects.

The final interaction, given by Eq. (VII-7c), varies as the second power of the applied magnetic field. This term is much easier to evaluate in the high field limit:

$$\sum (IM_I JM_J|\mathbf{P}^{(1)} \cdot \mathbf{H}|IM_I J'M_J')(IM_I J'M_J'|\mathbf{P}^{(1)} \cdot \mathbf{H}|IM_I JM_J'') \frac{1}{E_J - E_J'}$$

$$= \sum (J\|P^{(1)}\|J')^2 (-1)^{q+q'-M_J+M_{J'}} \begin{pmatrix} J & 1 & J' \\ -M_J & q & M_J' \end{pmatrix}$$

$$\times \begin{pmatrix} J' & 1 & J \\ -M_J' & q' & M_J'' \end{pmatrix} H_{-q} H'_{-q} \frac{1}{E_J - E_J'}$$

$$= \sum (-1)^{-Q+J+J'} [k]^{\frac{1}{2}} \begin{Bmatrix} 1 & 1 & k \\ J & J & J' \end{Bmatrix} (HH)^{(k)}_{-Q}$$

$$\times (J\|P^{(1)}\|J')^2 (-1)^{J-M_J} \begin{pmatrix} J & k & J \\ -M_J & Q & M_J'' \end{pmatrix} \frac{1}{E_J - E_J'}.$$

This can be written as a matrix element of the operator

$$\sum_k \mathbf{Z}^{(k)} \cdot (\mathbf{HH})^{(k)} \quad \text{(VII-11)}$$

where

$$(J\|Z^{(k)}\|J) = \sum_{J'} (-1)^{J+J'} \begin{Bmatrix} 1 & 1 & k \\ J & J & J' \end{Bmatrix} [k]^{\frac{1}{2}} \frac{(J\|P^{(1)}\|J')^2}{E_J - E_J'}.$$

Because of $(\mathbf{HH})^{(k)}$, the sum over k is limited to $k = 0$ or 2. The term with $k = 0$ will shift all M_J terms within a J multiplet by the same amount. The term with $k = 2$ will cause different M_J (or M_F) levels to shift differently, and should be considered when analyzing data.

These corrections are, in principle at least, considerably easier to evaluate than some of those discussed in previous sections. The added simplicity in this case arises because there is no sum over an infinite number of excited states; the sums involved are, rather, confined to the different states of a given configuration. This insures that the angular sums which occur in the second order expression can easily be carried out. In addition, the radial integrals which appear in the evaluation of these effects are the same as those which appear in the evaluation of the first order hyperfine interactions. Because of the rapidly decreasing size of the first order terms, the main second order effects can generally be assumed to arise from dipole-dipole and dipole-quadrupole interactions, thereby greatly limiting the number of terms which must be evaluated in the sums. We have not specified in our derivation whether the more accurate relativistic forms or the simpler non-relativistic forms of the operators should be used. Since these effects are second order, the non-relativistic expressions should be quite adequate in most cases. However, one may wish to keep the dipole interaction in its most general form (Eq. (IV-23)) in order to take into account both core polarization and relativity. One may, in fact, be able to use the second order terms to gain additional information as to the value of the different $\langle 1/r^3 \rangle$ terms, as we shall show in an example in the last chapter. Finally, let us note that the above equations also take into account the second-order breakdown of L and S by the hyperfine and Zeeman interactions. That is, the sum over intermediate states includes those states which differ in L and S from ψ_0.

If one wishes to evaluate the radial integrals which appear in the hyperfine calculation by fitting of experimental data, this can in principle be done if enough data exist by assuming that J is a good quantum number and fitting the data with the Hamiltonian

$$\sum_K (\mathbf{F}^{(K)} \cdot \mathbf{Q}^{(K)} + \mathbf{N}^{(K)} \cdot \mathbf{M}^{(K)} + \mathbf{G}_n^{(K)} \cdot \mathbf{H}_e^{(K)}) - g_J \mu_0 \mathbf{J} \cdot \mathbf{H} - g_I \mu_0 \mathbf{I} \cdot \mathbf{H}$$

$$+ \sum_{\kappa K} (\mathbf{A}_n^{(K)} \mathbf{T}_e^{(\kappa)})^{(1)} \cdot \mathbf{H} + \sum_K \mathbf{Z}^{(K)} \cdot (\mathbf{HH})^{(K)} \quad \text{(VII-12)}$$

which contains effective operators to correct for second order mixing of J states. After angular matrix elements have been evaluated, this will result in equations containing unknowns such as $\langle 1/r^3 \rangle_L$, $\langle 1/r^3 \rangle_Q$, and so on. If there are more equations than unknowns, these parameters can then be varied to obtain the best fit of the experimental data. In practice, however, one seldom has the computer programs necessary to fit energy levels with a Hamiltonian as complex as that given by (VII-12), although programs may well be available for the fitting with the first order terms alone. In such case, one can still easily obtain approximate corrections which can be improved by an iterative technique if necessary. Thus one calculates the unknown radial parameters ignoring the second order interactions, then uses these approximate values of the radial parameters to evaluate the second order corrections. The second order corrections can then be subtracted from the measured values and the remainder again fit using the first order terms. The new radial parameters obtained in this way can be used to calculate a better second order correction, etc. We shall consider such a calculation in more detail in Section IX-5.

4. ATOMIC ELECTRIC DIPOLE MOMENTS

Recently, a great deal of interest has arisen over the possibility of observing violations of time reversal invariance and parity in atomic systems.[161–169] One very sensitive indication of a violation of these invariance principles would be the measurement of a static electric dipole moment in an atom other than hydrogen. Since experimental investigations of this subject have been made in the hyperfine structure of the alkalis, these investigations are of interest in a study of hyperfine structure.

Earlier we stated that, due to parity conservation, there could be no static electric dipole moment of an atom. In fact, invariance of either parity or time reversal is sufficient to exclude electric dipole moments in stationary states of an atom. This can be shown following the work of Sandars.[166] Consider an atom in an external magnetic field **H**, with non-degenerate levels $|FM)$. If the atom possesses a static electric dipole moment, the interaction between the atom and an external electric field will take the form

$$H_s = -\mathbf{D} \cdot \mathbf{E}$$

where **D**, the electric dipole moment operator, is independent of **E**. We first show that if parity is conserved, that is, if the wavefunction $|FM)$ is an eigenvalue of the parity operator P, then **D** must vanish. The condition that the wavefunction be an eigenvalue of the parity operation implies that

$$P|FM) = (-1)^P|FM).$$

Conservation of parity also implies that H_s itself is invariant under the parity operation; since **E** changes sign under the parity operation, we must require

$$PDP^{-1} = -\mathbf{D}.$$

Then, since $P^{-1}P = P^\dagger P = 1$, we have

$$(FM|D_0|FM) = (FM|P^\dagger(PD_0 P^{-1})P|FM)$$
$$= -(FM|D_0|FM) = 0$$

which proves the first result.

Given the absence of a magnetic field, one can prove in a straightforward manner than an electric dipole moment is forbidden by time reversal invariance. The effect of the time reversal operator of an atom, T, on a state, $|FM\rangle$, that is an eigenvalue of a Hamiltonian which is invariant under time reversal, is simply[167]

$$T|FM\rangle = (-1)^{F-M+P}|F-M\rangle.$$

The Hamiltonian of a free atom is, of course, invariant under time reversal. Since the electric field is invariant under time reversal, overall invariance of H_s requires that

$$T\mathbf{D}T^{-1} = \mathbf{D}.$$

Finally, the time reversal operator T can be written as

$$T = KU$$

where K is the complex conjugation operator, and U is a unitary operator. Then, since $\mathbf{D}$ must be hermitian if it is to correspond to an observable:

$$(FM|D_0|FM) = ((FM|D_0|FM))^*$$
$$= (FM|K^\dagger(KD_0K^{-1})K|FM)$$
$$= (FM|T^\dagger(TD_0T^{-1})T|FM)$$
$$= (F-M|D_0|F-M).$$

Using the Wigner-Eckart theorem on the result above, one obtains

$$(-1)^{F-M}\begin{pmatrix}F & 1 & F\\-M & 0 & M\end{pmatrix}(F\|D\|F) = (-1)^{F+M}\begin{pmatrix}F & 1 & F\\M & 0 & -M\end{pmatrix}(F\|D\|F)$$
$$= -\left[(-1)^{F-M}\begin{pmatrix}F & 1 & F\\-M & 0 & M\end{pmatrix}(F\|D\|F)\right].$$

Because all 3-j symbols of the type $\begin{pmatrix}F & 1 & F\\-M & 0 & M\end{pmatrix}$ cannot vanish, this result implies that $(F\|D\|F) = -(F\|D\|F) = 0$. (The proof given by Sandars[166] for this case is incomplete as he has not shown that his $c \neq 0$ or ∞ (his Eq. 2.10). Thus, his Eq. 2.14 does not necessarily lead to the conclusion he has drawn.)

If a magnetic field is present, the total atomic Hamiltonian no longer commutes with the time reversal operator, and the proof given above is not valid. In this case, one can introduce an additional operator, R, which rotates the atom by an angle π around the y axis. Now, if the interaction of the atom with a uniform magnetic field along the z axis is

$$H_M = -\mu_z H_z$$

invariance of H_M with respect to time reversal of both the field and the atom requires that
$$T\mu_z T^{-1} = -\mu_z$$
or
$$TH_M T^{-1} = -H_M.$$
Rotation of the atom by π, however, changes μ_z to $-\mu_z$,
$$R\mu_z R^{-1} = -\mu_z,$$
so that
$$RTH_M T^{-1}R^{-1} = H_M.$$
Since the Hamiltonian of the free atom is invariant under both rotations and time reversal, the total Hamiltonian in the presence of a magnetic field is invariant under the combined operation TR. If we denote by $|\gamma\rangle$ an eigenvalue of the free atom plus magnetic field Hamiltonian, then $TR|\gamma\rangle$ must be equal in energy to $|\gamma\rangle$. However, as we saw in Section VI-1, the magnetic field should remove all degeneracies (except for accidental degeneracies, which we shall ignore). Thus, at most,
$$RT|\gamma\rangle = e^{i\delta}|\gamma\rangle.$$
Then, proceeding as before and using $R^\dagger R = 1$
$$\begin{aligned}(\gamma|D_0|\gamma) &= (\gamma|D_0|\gamma)^* \\ &= (\gamma|K^\dagger(KD_0 K^{-1})K|\gamma) \\ &= (\gamma|T^\dagger R^\dagger(RTD_0 T^{-1}R^{-1})RT|\gamma) \\ &= (\gamma|RTD_0 T^{-1}R^{-1}|\gamma).\end{aligned}$$
Since
$$RTD_0 T^{-1}R^{-1} = RD_0 R^{-1} = -D_0,$$
the above result implies that $D = 0$. This completes the proof that either time reversal invariance or conservation of parity is sufficient to forbid the existence of a static electric dipole moment in an atom.

Let us now consider a specific model which breaks these conservation laws, and investigate the effects of such a breakdown. In particular, let us consider a model in which the atomic electrons themselves possess electric dipole moments of magnitude d_e. Salpeter[168] has suggested that in such a case, a perturbation
$$H_d = -d_e \beta \sum_j (\boldsymbol{\sigma}_j \cdot \mathbf{E}_j + i\boldsymbol{\alpha}_j \cdot \mathbf{H}_j)$$
should be added to the free atom Hamiltonian, where $\mathbf{E}_j$ and $\mathbf{H}_j$ are the electric and magnetic fields at the jth electron. The value of d_e can be expected

to be very small from models of symmetry breaking ($d_e \sim 10^{-25}$ excm). Thus matrix elements of H_d must be very small and this interaction can be treated using perturbation theory.

The overall effect of the part of H_d caused by electric and magnetic fields internal to the atom is to admix very small components of opposite parity into the unperturbed state. In a complex atom, the effects of a very small admixture of this type are very difficult to measure. For instance, one effect would be to make possible electric dipole transitions between states which, in the absence of this perturbation, have the same parity. However, because of the many lines which appear in the spectrum of a complex atom, the difficulty in predicting transition wavelengths to a high accuracy, and the extreme weakness of a line which is allowed only by the breakdown of parity and time reversal, it would be virtually impossible to unambiguously identify a line as being caused by an effect of this type. (See, however, Salpeter[168] for a discussion of possible observable effects in very simple atoms.)

Thus, to find a possibly measurable effect, one must investigate the application of external fields to the atom. We consider first an external electric field. We are interested in interaction terms linear in the external electric field, as these will be interpreted as a static electric dipole moment of the atom. Such terms appear in both first and second order perturbation theory: in first order, one has $\langle -d_e \beta \boldsymbol{\sigma} \cdot \mathbf{E}_{\text{ext}} \rangle$, and in second order, $\langle -d_e \beta \boldsymbol{\sigma} \cdot \mathbf{E}_{\text{int}} \rangle \langle e\mathbf{r} \cdot \mathbf{E}_{\text{ext}} \rangle$. There is also a term of the type $\langle -d_e \beta i \boldsymbol{\alpha} \cdot \mathbf{H}_{\text{int}} \rangle \langle e\mathbf{r} \cdot \mathbf{E}_{\text{ext}} \rangle$ in second order; however, internal magnetic fields due to hyperfine interactions and the magnetic fine-structure interactions (spin-other-orbit, etc.) are much smaller than the electric central field of the atom. We therefore expect that the purely electrostatic terms above will dominate the magnetic field interaction.

With this approximation, the dipole moment has the expectation value valid to second order in perturbation theory

$$(\psi_0|D_0|\psi_0) = d_e(\psi_0|\beta\sigma_z|\psi_0) + \sum_i \frac{d_e}{E_0 - E_i}[(\psi_0|-\beta\boldsymbol{\sigma}\cdot\mathbf{E}_{\text{int}}|\psi_i) \\ \times (\psi_i|-ez|\psi_0) + (\psi_0|-ez|\psi_i)(\psi_i|-\beta\boldsymbol{\sigma}\cdot\mathbf{E}_{\text{int}}|\psi_0)].$$

Following Sandars,[166] we evaluate matrix elements of **D** for the special case of one-electron atoms, and we make the assumption that the hyperfine interactions do not contribute to $\mathbf{E}_{\text{int}}$. In such a case, **D** is a tensor of rank one acting in the space of the electrons only. For simplicity, we can evaluate D_0 in the state $\psi_0 = |II, jj\rangle$; matrix elements evaluated in this state can be easily related to matrix elements in any other state. Then

$$(\psi_0|D_0|\psi_0) = d_e(jj|\beta\sigma_z|jj) + \sum_i \frac{d_e}{E_0 - E_i}[(jj|-\beta\boldsymbol{\sigma}\cdot\mathbf{E}_{\text{int}}|\psi_i) \\ \times (\psi_i|-ez|jj) + (jj|-ez|\psi_i)(\psi_i|-\beta\boldsymbol{\sigma}\cdot\mathbf{E}_{\text{int}}|jj)]. \quad \text{(VII-13)}$$

154 Higher-Order Effects

In this special case, the matrix element of D_0 is independent of the nuclear state. However, in the usual case for which $\psi_0 = |IJFM\rangle$, the matrix element of D_0 will depend on I and F through a 6-j symbol of the type

$$\begin{Bmatrix} F & 1 & F \\ J & I & J \end{Bmatrix}$$

which results from use of Eq. (II-26).

It is convenient to replace $\beta\boldsymbol{\sigma}$ by $\boldsymbol{\sigma} + (\beta - 1)\boldsymbol{\sigma}$; the second term will be a completely relativistic one as $(\beta - 1)$ picks out the small component of the radial wavefunction. We first consider the parts of Eq. (VII-13) which can be written in terms of $\boldsymbol{\sigma}$ only:

$$D_{nr} = d_e(jj|\sigma_z|jj) + \sum_i \frac{d_e}{E_0 - E_i} [(jj|-\boldsymbol{\sigma}\cdot\mathbf{E}_{\text{int}}|\psi_i)(\psi_i|-ez|jj)$$
$$+ (jj|-ez|\psi_i)(\psi_i|-\boldsymbol{\sigma}\cdot\mathbf{E}_{\text{int}}|jj)].$$

The central field Hamiltonian for a one-electron atom is

$$H_0 = \boldsymbol{\alpha}\cdot c\mathbf{p} + \beta mc^2 - eV(r);$$

because ∇ commutes with everything in this expression except $eV(r)$, we can write

$$-\boldsymbol{\sigma}\cdot\mathbf{E}_{\text{int}} = \boldsymbol{\sigma}\cdot\nabla V(r) = -\frac{1}{e}[\boldsymbol{\sigma}\cdot\nabla, H_0].$$

Making this substitution in the second order term in D_{nr} and expanding the commutator leads immediately to

$$D_{nr} = d_e(jj|\sigma_z|jj) + \frac{d_e}{e}\sum_i [(jj|\boldsymbol{\sigma}\cdot\nabla|\psi_i)(\psi_i|-ez|jj)$$
$$- (jj|-ez|\psi_i)(\psi_i|\boldsymbol{\sigma}\cdot\nabla|jj)].$$

We can now invoke closure to carry out the sum over i, and the second order part of D_{nr} becomes simply

$$-d_e(jj|[\boldsymbol{\sigma}\cdot\nabla, z]|jj) = -d_e(jj|\sigma_z|jj);$$

this cancels exactly the first order part of D_{nr}, making D_{nr} vanish identically for a one-electron atom.

Let us consider now the term

$$D_1 = d_e(jj|(\beta - 1)\sigma_z|jj).$$

This is easily evaluated using the techniques of Chapters II and III. Thus

$$D_1 = -2d_e \int G_{nlj}^2 \, dr \begin{pmatrix} j & 1 & j \\ -j & 0 & j \end{pmatrix} \begin{Bmatrix} j & 1 & j \\ \tfrac{1}{2} & \bar{l} & \tfrac{1}{2} \end{Bmatrix} [j](-1)^{l+\frac{1}{2}+j+1} \langle \tfrac{1}{2} \| \sigma \| \tfrac{1}{2} \rangle$$

$$= -2d_e \int G_{nlj}^2 \, dr \left[\frac{j(j+1) - \bar{l}(\bar{l}+1) + \tfrac{3}{4}}{j+1} \right].$$

Here $\bar{l}$ is the orbital momentum of the small component of the state $|jj\rangle$. The remaining term in the matrix element of D_0 is given by

$$D_2 = \sum_{\psi_i} \frac{d_e}{E_0 - E_i} [(jj|(1-\beta)\boldsymbol{\sigma} \cdot \mathbf{E}_{\text{int}}|\psi_i)(\psi_i|-ez|jj)$$

$$+ (jj|-ez|\psi_i)(\psi_i|(1-\beta)\boldsymbol{\sigma} \cdot \mathbf{E}_{\text{int}}|jj)].$$

This term obviously cannot be so easily evaluated due to the sum over intermediate states. One can evaluate this term by actually calculating all of the excited state wavefunctions ψ_i and carrying out the sum, as was done in a similar situation discussed in Section V-6. Sandars[166] suggests the use of a Sternheimer procedure (Section V-8). In this case, one can find a state $|p\rangle$ such that D_2 can be written as

$$D_2 = 2d_e(jj|(1-\beta)\boldsymbol{\sigma} \cdot \mathbf{E}_{\text{int}}|p)$$

where the function $|p\rangle$ can be expanded as

$$|p\rangle = \sum_{\psi_i} \frac{|\psi_i\rangle(\psi_i|-ez|jj)}{E_0 - E_i}$$

and satisfies the equation

$$(H_0 - E_0)|p\rangle = \sum \frac{(E_i - E_0)|\psi_i\rangle(\psi_i|-ez|jj)}{E_0 - E_i}$$

$$= ez|jj\rangle.$$

Then, finally, we can evaluate D_2 as

$$D_2 = 4 d_e \int G_{nlj} G_p \frac{\partial}{\partial r} V(r) \, dr$$

where G_p is the small component of the state $|p\rangle$.

Sandars has carried out calculations of $R = (D_1 + D_2)/d_e$ for two states of hydrogen[166] and the ground states of several alkalis.[169] For the 2s state of hydrogen, for example, he calculates a value of $R \sim 120$, corresponding to a large enhancement of the dipole moment of the electron. Values of R for the alkalies can also be of the order of 100. Such large enhancements of the electron electric dipole moment in atoms has made possible experiments which seek to detect such a moment via detection of a static electric dipole moment of an atom. Thus far, however, no such moment has been experimentally detected, leading to the conclusion that if one exists, it corresponds to an electric dipole moment of an electron which is smaller than $(2.5) \times 10^{-23}$ e cm.[164]

The interaction H_d indicates that there will also be an interaction between the electron electric dipole moment and an external magnetic field. Since, however, the atom already experiences a first order interaction with a magnetic field, and since the electric dipole moment is exceedingly small (if not zero), such an effect will be very difficult to observe. A much more interesting effect would be a second order interaction felt by an atom in combined electric and magnetic fields:

$$H_{em} = \sum_\psi \frac{d_e}{E_0 - E} [(jj|\, e\mathbf{r} \cdot \mathbf{E}_{ext}|\psi_k)(\psi_k|-i\beta\mathbf{\alpha} \cdot \mathbf{H}_{ext}|jj)$$
$$+ (jj|-i\beta\mathbf{\alpha} \cdot \mathbf{H}_{ext}|\psi_k)(\psi_k|\, e\mathbf{r} \cdot \mathbf{E}_{ext}|jj)].$$

This term, being linear in the electric field, will contribute to the static electric dipole moment of the atom. In addition, this contribution to the atomic electric dipole moment will depend linearly on the applied magnetic field, which should aid in experimental verification of the effect.

5. DIFFERENTIAL POLARIZABILITY

Differential polarizability results from those Stark effect interactions which involve nuclear operators. In the language of perturbation theory, these are third order terms involving two interactions with the external field and one with the hyperfine interaction. One can calculate these terms by expanding the atomic wavefunction in powers of the external field

$$\psi = \psi^{(0)} + \psi^{(1)} + \psi^{(2)} + \cdots$$

and using this wavefunction to evaluate $H_{hyp} = \sum_J \mathbf{A}^{(J)} \cdot \mathbf{B}^{(J)}$. The terms of interest will be those proportional to the second powers of the electric field, that is,

$$(\psi^{(2)}|H_{hyp}|\psi^{(0)}) + (\psi^{(0)}|H_{hyp}|\psi^{(2)}) + (\psi^{(1)}|H_{hyp}|\psi^{(1)}).$$

Such interactions are represented by graphs of the type

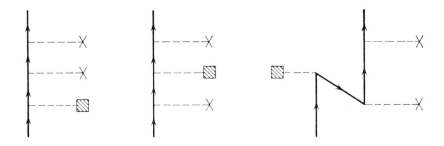

where X refers to an interaction with the electric field, ▨ to the hyperfine interaction.

Such calculations have been considered in detail by for example Schwartz,[170] Feichtner, et al.,[139] and Sandars.[140] We shall not discuss these more accurate calculations, but rather shall proceed from the viewpoint of effective operators. In this manner, we can determine the qualitative effects of such interactions. It is probably sufficient at this point to consider only the general form of these operators. This general form can easily be obtained through a straightforward argument. The two interactions with the external field must be expressable in terms of an effective operator of the type $\mathbf{P}_e^{(0k)k} \cdot (\mathbf{EE})^{(k)}$, where $k = 0, 2$. $\mathbf{P}_e^{(0k)}$ is a tensor in the space of the electrons which may contain both one- and two-body parts. We combine this operator with a hyperfine interaction of the form $\mathbf{T}_e^{(SL)J} \cdot \mathbf{B}_n^{(J)}$, where $\mathbf{T}_e^{(SL)}$ acts in the space of the electrons, $\mathbf{B}_n^{(J)}$ in the space of the nucleus. The resulting effective operator must have the general form

$$H_{\text{eff}} = \sum_{\kappa K} F(J, S, L, k, \kappa, K)(\mathbf{B}_n^{(J)} \mathbf{U}_e^{(S\kappa)K})^{(k)} \cdot (\mathbf{EE})^{(k)} \qquad \text{(VII-14)}$$

where $\mathbf{U}_e^{(S\kappa)}$ is an operator acting in the space of the electrons which may have both one- and two-body parts. F is a parameter whose value depends on the indicated variables. Triangular rules must be satisfied by the triads L, k, κ and k, J, K. Because of the appearance of the operator $\mathbf{B}_n^{(J)}$ in H_{eff}, the strength of H_{eff} will depend on the hyperfine state of the atom.

This third-order term will almost certainly be considerably smaller in magnitude than the second-order terms of Section VI-3. Therefore, these effects will be detectable primarily in cases in which the second-order terms vanish. For example, there can be no tensor Stark effect in $s_{1/2}$ or $p_{1/2}$ atomic states. The remaining Stark term, the scalar polarizability, shifts all hyperfine levels equally; thus, any unequal, or differential, Stark splitting between hyperfine levels must arise from this third-order interaction.

Consider first an atom containing a single $s_{1/2}$ electron outside of closed shells. In this case, the excited states appearing in the third-order perturbation equations clearly belong to configurations different from the ground configuration, either by promotion of the s electron to a np state, or by promotion of a core electron to some excited state. In either case, we have the restrictions $\kappa = 0$, $K = S$, implying that $k = L$. One finds, therefore, that there will be both scalar ($k = 0$) and tensor ($k = 2$) differential polarizabilities for atoms in the $s_{1/2}$ state. The former term arises from perturbations involving the dipole contact interaction ($S = 1, L = 0$); the latter arises from interactions involving the spin-dipolar ($S = 1, L = 2, J = 1$), quadrupole ($S = 0, L = 2, J = 2$) and octupole contact ($S = 1, L = 2, J = 3$) hyperfine terms. Because of the small size of the octupole moment, this last contribution will probably seldom be significant. Clearly, the scalar term will be larger than the tensor term in this case as the former depends in magnitude on the dipole contact interaction, which is, of course, very large for an $s_{1/2}$ electron.

The tensor polarizability also vanishes for atoms in the $p_{1/2}$ state. In this case, one might expect the dominant contribution to H_{eff} to arise from perturbations involving the near-by $p_{3/2}$ state. Perturbations of this type can be evaluated more accurately than those involving promotion of an electron into excited configurations because, in the former case, the energy denominators and wavefunctions are probably reasonably well known. Perturbation described by graphs in which the hyperfine interaction is either highest or lowest in time can be expressed directly in terms of the scalar and tensor polarizabilities of Section VI-3. This type of perturbation can be written in the form

$$-\frac{1}{2}\sum_{\psi_i}\frac{1}{\Delta E}[\langle\psi_0|\alpha_k\mathbf{W}^{(0k)k}\cdot(\mathbf{EE})^{(k)}|\psi_i\rangle\langle\psi_i|\mathbf{T}_e^{(SL)J}\cdot\mathbf{B}_n^{(J)}|\psi_0\rangle + c.c.]$$

The resulting operator must, of course, be expressible in the general form of Eq. (VII-14). In the $p_{1/2}$ case, the limitations on the summations of Eq. (VII-14) become simply $K = 0, 1$; $\kappa = 0, 1, 2$. Clearly the number of F parameters needed to describe differential polarizability in the $p_{1/2}$ state is much greater than the corresponding number required in the $s_{1/2}$ state.

These differential polarizabilities, though small, do lead to effects large enough to be measured. For example, Haun and Zacharias[171] have observed these terms in Cs, where they studied electric-field-induced energy shifts between levels having different F values. The shift observed was found to be described by the formula $\Delta\nu = -2.29\ 10^{-6}E^2$ cps, where E is measured in V/cm. Shifts of the same type have been observed in hydrogen by Fortson, et al.[172] Carrico, et al.[173] have studied energy shifts between levels of Cs having the same F quantum number but differing by one in the value of M_F.

A glance at Eq. (VII-14) shows that the scalar differential polarizability cannot cause such shifts between Zeeman substates, but that the tensor differential polarizability can: because the tensor differential polarizability is described by a tensor of rank two in the F space, diagonal matrix elements of this polarizability will be proportional to $(3M_F^2 - F(F+1))$. As noted above, however, the tensor differential polarizability should be much smaller than the scalar differential polarizability. This conclusion was confirmed by Carrico et al.[173] who found that the shifts observed by them could be described by the formula $\Delta v = -1.27 \; 10^{-8} E^2$ cps.

VIII. Hyperfine Structure of the One-Electron Atom

1. HYPERFINE INTERACTION CONSTANTS IN ONE-ELECTRON ATOMS

The interaction constants of Chapter IV take on particularly simple form when the atom of interest has only one electron outside of closed shells. In such a case, one can work directly with the hyperfine interaction as described by Eq. (IV-16). We find

$$a_{lj} = \frac{2\mu_0 \mu_I}{I} \frac{l(l+1)}{j(j+1)} \frac{1}{(\kappa-1)} \frac{e}{2\mu_0} P_{jj} \qquad \text{(VIII-1)}$$

$$b_{lj} = e^2 Q \frac{(2j-1)}{2j+2} T_{jj} \qquad \text{(VIII-2)}$$

$$c_{lj} = \mu_0 \Omega \frac{(8l)(l-1)(l+1)(l+2)}{(2j+2)(2j+3)(2j+4)} \frac{e}{2\mu_0} \frac{1}{(\kappa-2)} U_{jj} \qquad \text{(VIII-3)}$$

where the radial integrals P_{jj}, T_{jj} and U_{jj} are defined in Section IV-3. The non-relativistic limits of these interactions can be simply obtained using the relations of Section IV-6; then

$$a_{lj} = \frac{2\mu_0 \mu_I}{I} \frac{l(l+1)}{j(j+1)} \left\langle \frac{1}{r^3} \right\rangle \qquad l > 0$$

$$a_{0\frac{1}{2}} = \frac{16\pi\mu_0 \mu_I}{3I} |\psi_0|^2 \qquad l = 0, j = \tfrac{1}{2} \qquad \text{(VIII-4)}$$

where ψ_0 is the atomic wavefunction evaluated at $r = 0$. In the same limit

$$b_{lj} = e^2 Q \frac{(2j-1)}{(2j+2)} \left\langle \frac{1}{r^3} \right\rangle \qquad \text{(VIII-5)}$$

and

$$c_{lj} = \mu_0 \Omega \left\langle \frac{1}{r^5} \right\rangle \frac{(8l)(l-1)(l+1)(l+2)}{(2j+2)(2j+3)(2j+4)} \quad l > 1$$

$$c_{1\frac{3}{2}} = \frac{4}{35} \mu_0 \Omega \left| \frac{R_{n1}}{r^2} \right|^2_{r=0} \quad l = 1, j = \tfrac{3}{2}.$$
(VIII-6)

2. RELATIVISTIC RADIAL INTEGRALS—CASIMIR FACTORS

Correct evaluation of the operators given by Eqs. (VIII-1)–(VIII-3) requires correct values of the radial integrals P_{jj}, T_{jj}, and U_{jj}. In principle these integrals should be calculated using some form of Hartree-Fock-Dirac equation. In practice, such a calculation may not exist. Casimir[15] and Racah[13] suggested that reasonable approximations to the desired radial integrals could be obtained by assuming that all important contributions to the integrals come from very small r. This assumption is based on the importance of the r^{-n} weighting term in the radial integrals which favors greatly the region near to the origin. At small r, the electron can be assumed to be inside of the closed electronic shells, and thus will see the charge of the bare nucleus alone. If we further assume that $|mc^2 - E| \ll Ze/r$, the radial wavefunctions of the electrons must satisfy the equations

$$\left(\frac{d}{dr} - \frac{\kappa}{r}\right) F = -\frac{1}{\hbar c}\left(2mc^2 + \frac{Ze^2}{r}\right) G$$

$$\left(\frac{d}{dr} + \frac{\kappa}{r}\right) G = \frac{1}{\hbar c}\left(\frac{Ze^2}{r}\right) F.$$
(VIII-7)

The wavefunctions F and G which satisfy these equations can be expressed in terms of Bessel functions:

$$F = C[\tfrac{1}{2} x J_{2\rho+1}(x) - (\rho + \kappa) J_{2\rho}(x)]$$

$$G = -C[\alpha Z J_{2\rho}(x)]$$
(VIII-8)

where $x = (8Zr/a_0)^{1/2}$, and $\rho = (\kappa^2 - \alpha^2 Z^2)^{1/2}$. The quantity a_0 is the first Bohr radius of hydrogen, $a_0 = \hbar^2/me^2$. It is often useful to have expressions for F and G valid near to the origin. The above functions can be expanded in powers of r, with the lowest term in the expansion being

$$F = -\frac{C(\rho + \kappa)}{\Gamma(2\rho + 1)} \left(\frac{2Zr}{a_0}\right)^\rho$$

$$G = -\frac{Z\alpha C}{\Gamma(2\rho + 1)} \left(\frac{2Zr}{a_0}\right)^\rho.$$
(VIII-9)

The normalization constant C was calculated by Casimir using the wavefunctions of Eq. (VIII-8) for small r, a WKB solution for intermediate r, and the rigorous solution for an electron moving in a coulombic field $Z_0 = 1 + z$ (z = degree of ionization) for large r. He found

$$C^2 = \frac{dE}{dn} \bigg/ (2a_0 Z R_y hc) \tag{VIII-10}$$

where $R_y = me^4/4\pi\hbar^3 c$ is the Rydberg constant and n is the principal quantum number. The energy of one electron atoms can be conveniently expressed in terms of an effective quantum number n^*:

$$E = -R_y hc \frac{(Z_0)^2}{n^{*2}}.$$

The effective quantum number is related to the principal quantum number by

$$n = n^* + \sigma$$

where σ is the quantum defect. Expressed as a function of these quantities,

$$C^2 = \frac{Z_0^2}{a_0 Z n^{*3}} \left(1 - \frac{d\sigma}{dn}\right). \tag{VIII-11}$$

Casimir also showed that the energy separation δ of the states with $j' = l + \frac{1}{2}$ and $j'' = l - \frac{1}{2}$ could be expressed as

$$\delta = \frac{(\rho' - \rho'' - 1)}{hc} \frac{dE}{dn} \tag{VIII-12}$$

where δ is measured in cm^{-1} and

$$\rho' = ((l+1)^2 - Z^2\alpha^2)^{1/2}$$
$$\rho'' = (l^2 - Z^2\alpha^2)^{1/2}.$$

Using these quantities, we find that for $l > 0$,

$$C^2 = \frac{\delta}{2a_0 Z R_y (\rho' - \rho'' - 1)}. \tag{VIII-13}$$

Actually, the normalization constant for the $j' = l + \frac{1}{2}$ state (C') is slightly different in magnitude from that for the $j'' = l - \frac{1}{2}$ state (C''). Casimir found that

$$C' \simeq -C''$$

$$|C'|^2 - |C''|^2 = \frac{\alpha}{2ZR_y hc} \frac{d^2 E}{dn^2} (\rho' - \rho'' - 1). \tag{VIII-14}$$

Using the expressions for F and G given in Eq. (VIII-8), one can evaluate the integrals needed to calculate a, b, and c:[23,15]

$$\frac{e}{\mu_0(\kappa-1)} \int_0^\infty \frac{F_{nlj} G_{nlj}}{r^2} dr$$

$$= C^2 \left(\frac{2Z}{a_0}\right)^2 \frac{F}{(2l+1)(2l+2)(l)}$$

$$\int_0^\infty \frac{F_{nlj}^2 + G_{nlj}^2}{r^3} dr$$

$$= C^2 \left(\frac{2Z}{a_0}\right)^2 \frac{R}{l(2l+1)(2l+2)}$$

$$\frac{e}{\mu_0(\kappa-2)} \int_0^\infty \frac{F_{nlj} G_{nlj}}{r^4} dr$$

$$= C^2 \left(\frac{2Z}{a_0}\right)^4 \frac{10T}{(2l+3)(2l+2)(2l+1)(2l)(2l-1)(l+2)(2l-2)}$$

$$\frac{e}{\mu_0} \int_0^\infty \frac{F_{nlj'} G_{nlj''} + F_{nlj''} G_{nlj'}}{r^2} dr$$

$$= C'C'' \left(\frac{2Z}{a_0}\right)^2 \frac{G}{l(2l+1)(2l+2)}$$

$$\int_0^\infty \frac{F_{nlj'} F_{nlj''} + G_{nlj'} G_{nlj''}}{r^3} dr$$

$$= -C'C'' \left(\frac{2Z}{a_0}\right)^2 \frac{S}{l(2l+1)(2l+2)} \qquad \text{(VIII-15)}$$

where

$$R = \frac{l(l+1)(2l+1)}{\rho(\rho^2-1)(4\rho^2-1)} [3\kappa(\kappa+1) - \rho^2 + 1]$$

$$F = \frac{(2j)(2j+1)(j+1)}{\rho(4\rho^2-1)}$$

$$T = \frac{(2j+4)!(2\rho-4)!}{(2j-3)!(2\rho+3)!}$$

$$G = \frac{2l(l+1)\sin\pi(\rho'-\rho''-1)}{\pi Z^2 \alpha^2}$$

$$S = \frac{2l(l+1)\sin\pi(\rho' - \rho'' - 1)}{\pi Z^2 \alpha^2} \left\{ \frac{3(\rho' + l + 1)}{8(1 + \rho'') - 4l - 2} + \frac{3(\rho'' - l)}{8(1 + \rho' + 4l + 2)} \right.$$
$$\left. - \frac{1}{2}\frac{(\rho' + \rho'' - 1)}{(\rho' + \rho'' + 2)} - \frac{6[Z^2\alpha^2 + (\rho' + l + 1)(\rho'' - l)]}{3(2l+3)(2l-1) - 16Z^2\alpha^2} \right\}.$$

The above result is, of course, valid only when the orbital electron is completely inside of the closed atomic shells. This is, in general, too strong a requirement to place on the orbital electron. As in Section VII-2, we can hope to improve somewhat on this result if we divide these results by the nonrelativistic $\langle 1/r^n \rangle$ values calculated for an electron moving in the field of a nucleus of charge Z. The resulting relativistic correction factor should be more accurate than the total relativistic integral. We can calculate $\langle 1/r^3 \rangle$ by considering the fine structure splitting of the states j' and j''. For a nonrelativistic hydrogenic atom, one can use the well known result

$$\delta = (2l + 1) \frac{\mu_0^2 Z}{hc} \left\langle \frac{1}{r^3} \right\rangle \tag{VIII-16a}$$

where δ is measured in cm^{-1}. Casimir has shown that the relativistic fine structure splitting is just given by

$$\delta = (2l + 1) \frac{\mu_0^2 Z}{hc} \left\langle \frac{1}{r^3} \right\rangle H \tag{VIII-16b}$$

where

$$H = \frac{(\rho' - \rho'' - 1)(2l)(l+1)}{Z^2 \alpha^2}$$

and $H \to 1$ in the non-relativistic limit. Using Eqs. (VIII-16b) and (VIII-13), we find

$$\left\langle \frac{1}{r^3} \right\rangle = \frac{2Z^2 C^2}{a_0^2 l(l+1)(2l+1)}. \tag{VIII-17}$$

Using this result, Eqs. (VIII-15) can be written

$$\frac{e}{\mu_0} \frac{1}{(\kappa - 1)} \int \frac{F_{nlj} G_{nlj}}{r^2} dr = \left\langle \frac{1}{r^3} \right\rangle F$$

$$\int \frac{F_{nlj}^2 + G_{nlj}^2}{r^2} dr = \left\langle \frac{1}{r^3} \right\rangle R$$

$$-\frac{e}{\mu_0} \int \frac{F_{nlj'} G_{nlj''} + F_{nlj''} G_{nlj'}}{r^2} dr = \left\langle \frac{1}{r^3} \right\rangle G$$

$$\int \frac{F_{nlj'} F_{nlj''} + G_{nlj'} G_{nlj''}}{r^2} dr = \left\langle \frac{1}{r^3} \right\rangle S. \tag{VIII-18a}$$

In the same limit, we find[23]

$$\frac{e}{\mu_0}\frac{1}{(\kappa-2)}\int\frac{F_{nlj}G_{nlj}}{r^4}dr = \left\langle\frac{1}{r^5}\right\rangle T \quad \text{(VIII-18b)}$$

where

$$\left\langle\frac{1}{r^5}\right\rangle = C^2\left(\frac{2Z}{a_0}\right)^4 \frac{10}{(2l+3)(2l+2)(2l+1)(2l)(2l-1)(2l-2)(l+2)}$$

is the non-relativistic expectation value of $1/r^5$. Eqs. VIII-18 indicate that the relativistic integrals can be evaluated in an approximate fashion by using Hartree-Fock[174] values for $\langle 1/r^3 \rangle$, etc., in conjunction with the Casimir correction factors F, R, G, S, and T of Eqs. VIII-15. Radial integrals evaluated in this way should be considerably more accurate than those calculated using Eqs. VIII-15. Numerical values of the relativistic correction factors F, R, etc. can be obtained from tables given by Kopfermann.[134]

The accuracy of the Casimir factors is rather difficult to assess due to lack of correctly calculated relativistic integrals which could provide a standard of comparison. It is known, however, that relativistic effects cause contraction of s and p shells, and expansion of d, f, and so on shells. Thus we would expect that the effects of shielding might be very important for electrons with $l > 1$; this implies, of course, that the Casimir factors for $l > 1$ may be significantly in error. The accuracy of the p electron Casimir factors has been discussed by Schwartz.[175] He found that $F_{3/2}$ and $R_{3/2}$ are both overestimated by the Casimir approach; however, the error is rather small for low Z, increasing to the order of 10% for Tl ($Z = 81$).

3. APPROXIMATE HYPERFINE INTERACTION CONSTANTS

The results of the previous two sections can be combined to give approximate constants for the one electron atom:

$$a_{0\frac{1}{2}} = \frac{16\mu_0\mu_I}{3I}\frac{Z_0^2 Z}{a_0^3 n^{*3}}\left(1-\frac{d\sigma}{dn}\right)F$$

$$a_{lj} = \frac{2\mu_0\mu_I}{I}\frac{l(l+1)}{j(j+1)}\left\langle\frac{1}{r^3}\right\rangle F \quad (l>0)$$

$$b_{lj} = e^2 Q \frac{(2j-1)}{(2j+2)}\left\langle\frac{1}{r^3}\right\rangle R \quad \text{(VIII-19)}$$

$$c_{1\frac{3}{2}} = \frac{16}{315}\Omega\mu_0 \frac{Z^3 Z_0^2}{a_0^5 n^{*3}}\left(1-\frac{d\sigma}{dn}\right)T$$

$$c_{lj} = \mu_0 \Omega \frac{(8l)(l-1)(l+1)(l+2)}{(2j+2)(2j+3)(2j+4)}\left\langle\frac{1}{r^5}\right\rangle T \quad (l>1).$$

The expectation values of $1/r^3$ and $1/r^5$ can be related to the fine structure splittings through Eqs. (VIII-16) and (VIII-17), so that the above equations can be written as

$$a_{lj} = \frac{2\mu_0 \mu_I}{I} \frac{l(l+1)}{j(j+1)(2l+1)} \frac{\delta hcF}{\mu_0^2 ZH} \quad (l > 0)$$

$$b_{lj} = e^2 Q \frac{(2j-1)\delta hcR}{(2j+2)(2l+1)\mu_0^2 ZH} \quad \text{(VIII-20)}$$

$$c_{lj} = \frac{\mu_0 \Omega 20(l)(l+1)\delta hcZT}{(j+1)(j+2)(2j+3)(2l+3)(2l+1)(2l-1)a_0^2 \mu_0^2 H} \quad (l > 1).$$

The approximate results of Eq. (VIII-20) and many of the results of the previous section could be improved if somehow one could take into account the shielding of the nuclear charge by the electrons in filled shells. This can be done in an approximate way by using Eqs. (VIII-11) and (VIII-13) to describe the nonrelativistic fine structure splitting:

$$\delta = \frac{Z^2 Z_0^2 R_y \alpha^2}{l(l+1)n^{*3}} \left(1 - \frac{d\sigma}{dn}\right).$$

If we assume $d\sigma/dn = 0$, the above relationship is simply the familiar Lande formula for the fine structure separation. It is known that this formula works remarkably well if the bare nuclear charge Z is replaced by Z_i, where $Z_i = Z - 4$ for p electrons, $Z_i = Z - 11$ for d electrons, and $Z_i = Z - 35$ for $4f$ electrons. Use of Z_i rather than Z in the results above should improve the accuracy of these results by partially taking into account shielding effects. Even with this correction, however, Eq. (VIII-20) must be considered to give only very rough results.

IX. Hyperfine Structure in the Many-Electron Atom

1. HYPERFINE STRUCTURE IN THE CONFIGURATION l^N

The evaluation of the hyperfine interaction constants can be separated into two steps: (1) evaluation of radial matrix elements; and (2) evaluation of angular matrix elements. The radial matrix elements (P_{++}, T_{++}, etc.) should be evaluated using radial wave-functions which result from some type of Hartree-Fock-Dirac calculation. It is often interesting, however, to treat the radial integrals as unknown parameters which can be varied in order to make calculated hyperfine constants agree with measured values. In this way, one takes into account not only relativity, but also most forms of core polarization.

The calculation of angular matrix elements reduces to the calculation of a number of matrix elements of the type $\langle JM_J | W^{(\kappa k)K}_\pi | J'M'_J \rangle$ (or, equivalently $(JM_J | R^{(\kappa k)K}_\pi | J'M'_J)$). In general, the basis states used to determine the zeroth order state $|JM_J\rangle$ are of the form $|l^N \gamma SLJM\rangle$ where γ stands for additional quantum numbers needed to specify the state. Thus, as a result of the fitting procedure of Section III-10 we must have

$$|JM_J\rangle = \sum_{\gamma SL} a(\gamma, S, L) | l^N \gamma SLJM_J \rangle.$$

Assuming that the $a(\gamma, S, L)$'s are known, our problem becomes simply that of evaluating matrix elements of the type

$$\langle l^N \gamma SLJM_J | W^{(\kappa k)K}_\pi | l^N \gamma' S'L'J'M'_J \rangle.$$

Then, using Eqs. (II-19) and (II-23)

$$\langle l^N\gamma SLJM_J| W^{(\kappa k)K}_\pi |l^N\gamma' S'L'J'M'_J\rangle = (-1)^{J-M_J}[J,K,J']^{1/2}$$

$$\times \begin{pmatrix} J & K & J' \\ -M_J & \pi & M'_J \end{pmatrix} \begin{Bmatrix} S & S' & \kappa \\ L & L' & k \\ J & J' & K \end{Bmatrix} \langle l^N\gamma SL\|W^{(\kappa k)}\|l^N\gamma'S'L'\rangle. \quad \text{(IX-1)}$$

The reduced matrix element above can be calculated using the equation

$$\langle l^N\gamma SL\|W^{(\kappa k)}\|l^N\gamma'S'L'\rangle = N[S,S',\kappa,L,L',k]^{1/2}$$

$$\times \sum_{\bar\gamma\bar S\bar L}(l^N\gamma SL\{|l^{N-1}\bar\gamma\bar S\bar L)(l^N\gamma'S'L'\{|l^{N-1}\bar\gamma\bar S\bar L)$$

$$\times (-1)^{S+1/2+\bar S+\kappa+\bar L+l+L+k}\begin{Bmatrix}S & \kappa & S' \\ \tfrac12 & \bar S & \tfrac12\end{Bmatrix}\begin{Bmatrix}L & k & L' \\ l & \bar L & l\end{Bmatrix}$$

(IX-2)

where $(l^N\gamma SL\{|l^{N-1}\bar\gamma\bar S\bar L)$ is a coefficient of fractional parentage (CFP). Tables of CFP have been given by Nielson and Koster[176] for states of the p^n, d^n, and f^n configurations.

Certain matrix elements can be immediately shown to vanish using selection rules based on the groups $Sp(4l+2)$, $R(2l+1)$, G_2, and $R(3)$. The latter group provides, of course, the familiar selection rules on angular momentum which require that the matrix element in Eq. (IX-1) vanish unless $|J-J'|\leq K\leq J+J'$, $|S-S'|\leq \kappa\leq S+S'$, and $|L-L'|\leq k\leq L+L'$. The selection rules based on the group $Sp(4l+2)$ can easily be obtained by considering the equivalent problem of selection rules based on quasispin. Operators $\mathbf{W}^{(\kappa k)}$ where $\kappa+k$ is odd have quasispin rank zero, which implies that matrix elements of $\mathbf{W}^{(\kappa k)}$ ($\kappa+k$ odd) must be diagonal in quasispin or, equivalently, seniority. On the other hand, operators $\mathbf{W}^{(\kappa k)}$ where $\kappa+k$ is even have quasispin rank one; these operators can therefore couple states differing in quasispin by zero or one, which corresponds to a difference in seniority of zero or two. Thus, matrix elements of the dipole and octupole interactions are diagonal in seniority, whereas those of the quadrupole interaction may differ in seniority by 0 or 2.

The selection rules based on the groups $R(2l+1)$ and G_2 cannot be so concisely stated. However, selection rules based on the groups $R(7)$ (f electrons) and $R(5)$ (d electrons) can be obtained with the help of Tables (IX-1)–(IX-4) if we recall that the $\mathbf{W}_\pi^{(\kappa k)}$ (k odd) form the components of a tensor transforming like (110) in $R(7)$, (11) in $R(5)$; and that the $\mathbf{W}_\pi^{(\kappa k)}$ (k even) form the components of a tensor transforming like (200) in $R(7)$, or (20) in $R(5)$. For example, Table (IX-1) lists the coefficients $c(WW'(110))$; if $c(WW'(110)) = 0$, the operator $\mathbf{W}^{(\kappa k)}$ with k odd cannot connect an atomic state transforming like W in $R(7)$ to an atomic state transforming like W' in $R(7)$. Likewise, Table (IX-2) lists the coefficients $c(WW'(200))$ needed for f electron calculations; and Tables (IX-3) and (IX-4) list the coefficients $c(WW'(20))$ and $c(WW'(11))$ used in d electron calculations. Finally, selection rules for G_2 (f electrons) can be obtained using Tables (IX-5)–(IX-7) by recalling that $\mathbf{W}_\pi^{(\kappa 2)}$, $\mathbf{W}_\pi^{(\kappa 4)}$, and $\mathbf{W}_\pi^{(\kappa 6)}$ form the components of a tensor transforming like (20) of G_2; $\mathbf{W}_\pi^{(\kappa 1)}$ and $\mathbf{W}_\pi^{(\kappa 5)}$, a tensor transforming like (11) of G_2; and $\mathbf{W}_\pi^{(\kappa 3)}$, a tensor transforming like (10) of G_2.

Table IX-1. The Coefficients $C(WW'(110))$

$W \backslash W'$	(000)	(100)	(110)	(200)	(111)	(210)	(211)	(220)	(221)	(222)
(000)			1							
(100)		1			1	1				
(110)	1		1	1	1		1	1		
(200)			1	1			1			
(111)		1	1		1	1	1		1	
(210)		1			1	2	1		1	
(211)			1	1	1	1	2	1	1	1
(220)			1				1	1	1	
(221)					1	1	1	1	2	1
(222)							1		1	1

Table IX-2. The Coefficients $C(WW'(200))$

$W \backslash W'$	(000)	(100)	(110)	(200)	(111)	(210)	(211)	(220)	(221)	(222)
(000)				1						
(100)		1				1				
(110)			1	1			1			
(200)	1		1	1				1		
(111)					1	1	1			
(210)		1			1	2			1	
(211)			1			1	2	1	1	
(220)				1			1	1		1
(221)					1	1			2	1
(222)								1	1	1

Table IX-3. The Coefficients $C(WW'(20))$

$W \backslash W'$	(00)	(10)	(11)	(20)	(21)	(22)
(00)				1		
(10)		1			1	
(11)			1	1	1	
(20)	1		1	1		1
(21)		1	1		2	1
(22)				1	1	1

Table IX-4. The Coefficients $C(WW'(11))$

$W \backslash W'$	(00)	(10)	(11)	(20)	(21)	(22)
(00)			1			
(10)		1	1		1	
(11)	1	1	1	1	1	1
(20)			1	1	1	
(21)		1	1	1	2	1
(22)			1		1	1

Table IX-5. The Coefficients $C(UU'(20))$

$U \backslash U'$	(00)	(10)	(11)	(20)	(21)	(30)	(22)	(31)	(40)
(00)				1					
(10)		1	1	1	1	1			
(11)		1	1	1	1	1		1	
(20)	1	1	1	2	2	1	1	1	1
(21)		1	1	2	2	2	1	2	1
(30)		1	1	1	2	2	1	2	1
(22)				1	1	1	1	1	1
(31)			1	1	2	2	1	3	2
(40)				1	1	1	1	2	2

Table IX-6. The Coefficients $C(UU'(11))$

$U \backslash U'$	(00)	(10)	(11)	(20)	(21)	(30)	(22)	(31)	(40)
(00)			1						
(10)		1		1	1				
(11)	1			1	1		1	1	
(20)		1	1	1	1	1		1	
(21)		1	1	1	2	1		1	1
(30)			1	1	1	1	1	1	1
(22)			1			1	1	1	
(31)				1	1	1	1	2	1
(40)					1	1		1	1

Table IX-7. The Coefficients $C(UU'(10))$

$U \backslash U'$	(00)	(10)	(11)	(20)	(21)	(30)	(22)	(31)	(40)
(00)		1							
(10)	1	1	1	1					
(11)		1		1	1				
(20)		1	1	1	1	1			
(21)			1	1	1	1	1	1	
(30)				1	1	1		1	1
(22)					1			1	
(31)					1	1	1	1	1
(40)						1		1	1

Even stronger selection rules can be obtained for the operators $\mathbf{W}^{(0k)}$ with k odd, which form the generators of $R(2l+1)$. These operators can link only states transforming according to the same representations W of $R(2l+1)$. Likewise, $\mathbf{W}^{(05)}$ and $\mathbf{W}^{(01)}$ form the generators of G_2, and therefore can connect only states transforming according to the same representation of G_2.

Matrix elements of $\mathbf{W}^{(01)}$ and $\mathbf{W}^{(10)}$ can easily be calculated by exploiting the proportionality of these operators to the operators $\mathbf{L}$ and $\mathbf{S}$ (Section IV-4). Matrix elements of $\mathbf{L}$ and $\mathbf{S}$ diagonal in J are simply obtained using the reduced matrix elements

$$\langle l^N \gamma SLJ \| \mathbf{L} \| l^N \gamma' S'L'J \rangle$$
$$= \delta(\gamma, \gamma')\delta(S, S')\delta(L, L') \frac{[J(J+1) + L(L+1) - S(S+1)]}{2J(J+1)}$$
$$\times [J(J+1)(2J+1)]^{1/2} \quad \text{(IX-3)}$$

and

$$\langle l^N\gamma SLJ\|S\|l^N\gamma'S'L'J\rangle$$
$$= \delta(\gamma,\gamma')\delta(S,S')\delta(L,L')\frac{[J(J+1)+S(S+1)-L(L+1)]}{2J(J+1)}$$
$$\times [J(J+1)(2J+1)]^{1/2}$$

Matrix elements off-diagonal in J can be obtained using Eqs. (II-25) and (II-26) and the reduced matrix elements

$$\langle L\|L\|L\rangle = [L(L+1)(2L+1)]^{1/2}$$

and

$$\langle S\|S\|S\rangle = [S(S+1)(2S+1)]^{1/2}.$$

One finds

$$\langle l^N\gamma SLJ\|L\|l^N\gamma'S'L'J'\rangle$$
$$= \delta(\gamma,\gamma')\delta(S,S')\delta(L,L')(-1)^{J-J_>+1}$$
$$\times \left[\frac{(S+L+J_>+1)(L-S+J_>)(J_>+S-L)(S+L-J_>+1)}{4J_>}\right]^{1/2}$$
$$= -\langle l^N\gamma SLJ\|S\|l^N\gamma'S'L'J'\rangle \tag{IX-4}$$

where $J_>$ is the larger of J, J'.

Neilson and Koster[176] have also tabulated, for the states of p^n, d^n, and f^n configurations ($n \leq 2l+1$), reduced matrix elements of the operators $\mathbf{V}^{11}$ ($\mathbf{V}^{11} = \mathbf{W}^{(11)}/\sqrt{6}$) and $\mathbf{U}^k$ ($\mathbf{U}^k = (2/[k])^{1/2}\mathbf{W}^{(0k)}$). These tables can, of course, be used directly to evaluate needed matrix elements $\langle l^N\gamma SL\|W^{(0k)}\|l^N\gamma'S'L'\rangle$. They can also be used indirectly to obtain most of the remaining matrix elements which may be needed using an equation obtained by Judd:[177]

$$\langle l^{Na}v_1\gamma S_1L\|W^{(1k)}\|l^{Na}v_1\gamma'S_1'L'\rangle$$
$$= (-1)^{x+2l+1+(N_b+v_2)/2}(2l+2-v_1)^{-1/2}$$
$$\times \begin{pmatrix} \frac{1}{2}(2l+1-v_2) & 1 & \frac{1}{2}(2l+1-v_2') \\ \frac{1}{2}(2l+1-N_b) & 0 & -\frac{1}{2}(2l+1-N_b) \end{pmatrix}^{-1}$$
$$\times \langle l^{Nb}v_2\gamma S_2L\|W^{(0k)}\|l^{Nb}v_2'\gamma'S_2L'\rangle \tag{IX-5}$$

for k even, where

$$S_2 = \tfrac{1}{2}(2l+1-v_1)$$
$$v_2 = 2l+1-2S_1$$
$$v_2' = 2l+1-2S_1'.$$

Judd found for f electrons,

$$x = v_1 + 1 + v_2\,\delta(v_2, v_2').$$

One can show that x satisfies the same relationship in the case of d electrons. Here v_1, v_2, and so on, are seniorities, and N_b can have any value for which the states $|l^{N_b}v_2\,\gamma S_2\,L\rangle$ and $|l^{N_b}v_2'\,\gamma S_2\,L'\rangle$ exist.

Matrix elements for states with $n > 2l + 1$ can be obtained from these tables by utilizing the Wigner-Eckart theorem in quasispin space. Using the relationship of quasispin to seniority and Eq. (II-19), we find

$$\langle l^N v\gamma SLM_S\,M_L|W_{\pi q}^{(\kappa k)}|l^N v'\gamma' S'L'M_S'\,M_L'\rangle$$

$$= \langle QM_Q\gamma SLM_S\,M_L|W_{\pi q}^{(\kappa k)}|Q'M_Q\gamma' S'L'M_S'\,M_L'\rangle$$

$$= (-1)^{\bar Q - M_Q}\begin{pmatrix} Q & \bar Q & Q' \\ -M_Q & 0 & M_Q \end{pmatrix} A \qquad (\kappa + k \neq 0) \qquad \text{(IX-6)}$$

where

$$Q = \tfrac{1}{2}(2l + 1 - v)$$
$$Q' = \tfrac{1}{2}(2l + 1 - v')$$
$$M_Q = -\tfrac{1}{2}(2l + 1 - N),$$

and $\bar Q = 0$ or 1 as required to make $\bar Q + \kappa + k$ odd. A is a number independent of M_Q. A state having the same quantum numbers but belonging to the conjugate configuration l^{4l+2-N} will be an eigenfunction of Q_z with eigenvalue

$$M_Q' = -\tfrac{1}{2}(2l + 1 - 4l - 2 + N)$$
$$= \tfrac{1}{2}(2l + 1 - N) = -M_Q.$$

Using the Wigner-Eckart theorem in quasispin space again, we find

$$\langle l^{4l+2-N}v\gamma SLM_S\,M_L|W_{\pi q}^{(\kappa k)}|l^{4l+2-N}v'\gamma' S'L'M_S'\,M_L'\rangle$$

$$= (-1)^{\bar Q + M_Q}\begin{pmatrix} Q & \bar Q & Q' \\ M_Q & 0 & -M_Q \end{pmatrix} A \qquad \text{(IX-7)}$$

where A is the same function which appears in Eq. (IX-6). Combining these results, we find that reduced matrix elements for states in the second half of the shell can be obtained from those for states in the first half, using the relation

$$\langle l^N v\gamma SL\|W^{(\kappa k)}\|l^N v'\gamma' S'L'\rangle = (-1)^{\kappa + k + 1 + (v-v')/2}$$

$$\langle l^{4l+2-N}v\gamma SL\|W^{(\kappa k)}\|l^{4l+2-N}v'\gamma' S'L'\rangle \qquad (\kappa + k \neq 0). \qquad \text{(IX-8)}$$

It should be emphasized that this relationship holds only if states in the second half of the shell are defined in terms of particles (see e.g. Jucys,[178] Judd,[43]

Armstrong[175]). If states in the second half of the shell are to be defined in terms of holes, the above result should be multiplied by $(-1)^{(v'-v)/2}$ (see e.g. Racah[19]).

The matrix elements of $W^{(1k)}$ (k odd) have not been tabulated, and they must be obtained using Eqs. (IX-1) and (IX-2). Some easing of the task of calculation can often result, however, through use of a relationship based on the result of Eq. (IX-6):

$$\langle l^{N_a} v\gamma SL \| W^{(\kappa k)} \| l^{N_a} v'\gamma' S'L' \rangle$$
$$= \langle l^{N_b} v\gamma SL \| W^{(\kappa k)} \| l^{N_b} v'\gamma' S'L' \rangle (-1)^{(N_b - N_a)/2}$$
$$\times \frac{\begin{pmatrix} \frac{1}{2}(2l+1-v) & \bar{Q} & \frac{1}{2}(2l+1-v') \\ \frac{1}{2}(2l+1-N_a) & 0 & -\frac{1}{2}(2l+1-N_a) \end{pmatrix}}{\begin{pmatrix} \frac{1}{2}(2l+1-v) & \bar{Q} & \frac{1}{2}(2l+1-v') \\ \frac{1}{2}(2l+1-N_b) & 0 & -\frac{1}{2}(2l+1-N_b) \end{pmatrix}}. \quad \text{(IX-9)}$$

As above, $\bar{Q} = 0$ or 1 as required to make $\bar{Q} + \kappa + k$ odd. Eq. (IX-8) can be seen to be a special case of this result. Eq. (IX-9) can be very useful for example when $v = v'$, in which case this equation can relate a matrix element calculated in the configuration $N_b = v$ to a matrix element in the configuration $N_a > v$.

2. THE HALF-FILLED SHELL

The half-filled shell, $l^h (h = 2l + 1)$, has a property which can be very useful in the evaluation of matrix elements. This property is based on Eq. (IX-6). In the half-filled shell, $M_Q = 0$, and a matrix element of $W^{(\kappa k)}$ is proportional to the 3-j symbol

$$\begin{pmatrix} Q & \bar{Q} & Q' \\ 0 & 0 & 0 \end{pmatrix}$$

which vanishes unless $Q + \bar{Q} + Q'$ is even. As before, $\bar{Q} = 0$ or 1 as required to make $\kappa + k + \bar{Q}$ odd. We find, therefore, that in this case, the operators $W^{(\kappa k)}$ with $\kappa + k$ even can connect only states which differ by two in seniority. This means that the quadrupole interaction has no non-vanishing diagonal matrix elements (either relativistic or non-relativistic) in states of the half-filled shell. Note, however, that the pseudo-quadrupole term which results from the breakdown of J by the dipole interaction (Section VII-3) is not required to vanish, and one may therefore observe a "quadrupole" interaction in the half-filled shell even if L and S are good quantum numbers. In any case, if LS coupling holds, or if seniority is approximately a good quantum number, one can expect the quadrupole interaction to be small for states in the half-filled shell.

We must also note that the Hunds rule ground state for the half-filled shell is ^{h+1}S. The only hyperfine interaction which does not have zero expectation value for such a state is the contact term in the dipole interaction, which is relativistic in origin if $l > 0$. (There is also, of course, the possibility of a core polarization contribution to this term.) The ground state of an atom with a half-filled shell is therefore a very good place to study relativistic effects in hyperfine structure.

3. THE CONTACT INTERACTION IN l^n

One can use the Casimir factors of the previous chapter in order to obtain approximate values for the parameters $\langle \delta(r)/r^2 \rangle_{10}$;

$$\left\langle \frac{\delta(r)}{r^2} \right\rangle_{10} = \frac{1}{(2l+1)^2} [2(l+1)^2 l F_+ - 2l(l+1)G - 2l^2(l+1)F_-] \left\langle \frac{1}{r^3} \right\rangle,$$

where F_+ is the F of Eq. (VIII-15) evaluated for $j = l + \frac{1}{2}$, etc. We wish to expand F and G in powers of $(Z\alpha)^2$. Thus

$$F = \frac{2j(2j+1)(j+1)}{\rho(4\rho^2 - 1)} \simeq 1 + \frac{(Z\alpha)^2(6\kappa^2 - \frac{1}{2})}{\kappa^2(4\kappa^2 - 1)} + O((Z\alpha)^4) + \cdots$$

$$G = \frac{2l(l+1) \sin \pi(\rho' - \rho'' - 1)}{\pi(Z\alpha)^2} \simeq 1 - O((Z\alpha)^4) + \cdots.$$

With these approximations, one finds

$$\left\langle \frac{\delta(r)}{r^2} \right\rangle_{10} (s \text{ electrons}) = \frac{4ZZ_0^2}{n^{*3} a_0^3}$$

and

$$\left\langle \frac{\delta(r)}{r^2} \right\rangle_{10} (l > 0 \text{ electrons}) = -\frac{6(Z\alpha)^2(4l^2 + 4l - 1)}{l(l+1)(2l+1)^3(2l+3)(2l-1)} \frac{ZZ_0^2}{n^{*3} a_0^3}$$

where we have also used Casimir's $\langle 1/r^3 \rangle$ values. Thus for electrons in the same atom, one has the approximate ratio:

$$\frac{\langle \delta(r)/r^2 \rangle_{10}(l > 0 \text{ electrons})}{\langle \delta(r)/r^2 \rangle_{10} (s \text{ electrons})} \simeq -\frac{3(Z\alpha)^2 n_s^{*3}}{16 l^5 n_l^{*3}}. \quad \text{(IX-10)}$$

We note that $\langle \delta(r)/r^2 \rangle_{10}$ for $l > 0$ is of the opposite sign to that for $l = 0$. Thus the "relativistic" contact interaction for $l > 0$ differs in sign from the non-relativistic interaction, at least to order $(Z\alpha)^2$.

4. HYPERFINE STRUCTURE IN THE CONFIGURATION $l^N l'^M$

The results of this section will perhaps be most useful if we do not consider the hyperfine interactions directly, but rather concern ourselves with the interaction constants A, b, and c. We shall assume that the state of interest $|l^N l'^M \gamma JM\rangle$, can be expanded in a series of the type

$$|l^N l'^M \gamma JM\rangle = \sum c_{\gamma J}(\alpha J_1, \beta J_2)|l^N \alpha J_1, l'^M \beta J_2, \gamma JM\rangle$$

where the sum is over α, β, J_1, and J_2. The expansion coefficient $c_{\gamma J}(\alpha J_1; \beta J_2)$ can be determined from an analysis of the fine structure (Section III-10). Then

$$A = \frac{\mu_I}{IJ} \sum c_{\gamma J}(\alpha J_1; \beta J_2) c_{\gamma J}(\alpha' J'_1; \beta' J'_2)$$

$$\times (l^N \alpha J_1, l'^M \beta J_2; \gamma JJ | M^{(1)}_{D0} | l^N \alpha' J'_1, l'^M \beta' J'_2; \gamma JJ). \quad \text{(IX-11a)}$$

The matrix element above is easily evaluated using Eqs. (II-25) and (II-26):

$$(l^N \alpha J_1, l'^M \beta J_2; \gamma JJ | M^{(1)}_{D0} | l^N \alpha' J'_1, l'^M \beta' J'_2; \gamma JJ)$$

$$= (-1)^{J+1}[J]\begin{pmatrix} J & 1 & J \\ -J & 0 & J \end{pmatrix}$$

$$\times \left[(-1)^{J_1 + J_2} \begin{Bmatrix} J & 1 & J \\ J'_1 & J_2 & J_1 \end{Bmatrix} \delta(J_2, J'_2)\delta(\beta, \beta')(l^N \alpha J_1 \| M^{(1)}_D \| l^N \alpha' J'_1) \right.$$

$$\left. + (-1)^{J'_1 + J'_2} \begin{Bmatrix} J & 1 & J \\ J'_2 & J_1 & J_2 \end{Bmatrix} \delta(J_1, J'_1)\delta(\alpha, \alpha')(l'^M \beta J_2 \| M^{(1)}_D \| l'^M \beta' J'_2) \right].$$

(IX-11b)

In the special case that J_1 and J_2 are good quantum numbers (i.e. $c_{\gamma J}(\alpha J_\alpha; \beta J_\beta) = \delta(\alpha J_\alpha, \phi J_1)\delta(\beta J_\beta, \varepsilon J_2)$) this result takes on the simple form

$$A = \frac{1}{2J(J+1)} \{[J(J+1) + J_1(J_1+1) - J_2(J_2+1)]A(l^N \phi J_1)$$

$$+ [J(J+1) + J_2(J_2+1) - J_1(J_1+1)]A(l'^M \varepsilon J_2)\} \quad \text{(IX-12)}$$

where

$$A(l^N \phi J_1) = \frac{\mu_I}{I} \frac{(l^N \phi J_1 \| M^{(1)}_D \| l^N \phi J_1)}{[J_1(J_1+1)(2J_1+1)]^{\frac{1}{2}}}$$

is the dipole interaction constant for the state $|l^N\phi J_1\rangle$, and so on.

In the special case that

$$|\gamma JJ\rangle = |l^N\alpha J_1, s_{1/2}; \gamma JJ\rangle$$

one has the corresponding expression for A

$$A = \frac{1}{2J(J+1)}\{[J(J+1) + J_1(J_1+1) - \tfrac{1}{2}(\tfrac{3}{2})]A(l^N\alpha J_1)$$

$$+ [J(J+1) + \tfrac{1}{2}(\tfrac{3}{2}) - J_1(J_1+1)]A(s)\}. \quad \text{(IX-13)}$$

As can be shown using the approximate formulae of the previous chapter, $a_{0\,1/2}(= A(s))$ is usually considerably larger than a_{1j}, and consequently, except in special cases, much larger than $A(l^N\alpha J_1)$. However, as has been emphasized by Wybourne,[45] we cannot in general ignore the contribution of $A(l^N\alpha J_1)$ to Eq. (IX-13) even if that is true. This follows because the factor weighting the contribution of $A(l^N\alpha J_1)$ to A is approximately $2J$ as large as that weighting the contribution of $A(s)$ to A. Thus, if J is large, as it often is for rare-earth and actinide atoms, the contribution to A from the two shells may be comparable, even though $A(l^N\alpha J_1)$ is considerably smaller than $A(s)$.

In the same state of $l^N l'^M$, the quadrupole constant b is given by

$$b = 2eQ \sum c_{\gamma J}(\alpha J_1; \beta J_2) c_{\gamma J}(\alpha' J_1'; \beta' J_2')$$

$$(l^N\alpha J_1, l'^M\beta J_2; \gamma JJ| Q_{D0}^{(2)} |l^N\alpha' J_1', l'^M\beta' J_2'; \gamma JJ)$$

where

$$(l^N\alpha J_1, l'^M\beta J_2; \gamma JJ| Q_{D0}^{(2)} |l^N\alpha' J_1', l'^M\beta' J_2'; \gamma JJ)$$

$$= \begin{pmatrix} J & 2 & J \\ -J & 0 & J \end{pmatrix}[J](-1)^J$$

$$\times \Bigg[(-1)^{J_1+J_2}\begin{Bmatrix} J & 2 & J \\ J_1' & J_2 & J_1 \end{Bmatrix}\delta(J_2, J_2')\delta(\beta, \beta')(l^N\alpha J_1 \| Q_D^{(2)} \| l^N\alpha' J_1')$$

$$+ (-1)^{J_1'+J_2'}\begin{Bmatrix} J & 2 & J \\ J_2' & J_1 & J_2 \end{Bmatrix}\delta(J_1, J_1')\delta(\alpha, \alpha')(l'^M\beta J_2 \| Q_D^{(2)} \| l'^M\beta' J_2')\Bigg].$$

$$\text{(IX-14)}$$

Finally, the octupole constant c can be obtained from the relationship

$$c = -\Omega \sum c_{\gamma J}(\alpha J_1; \beta J_2) c_{\gamma J}(\alpha' J_1'; \beta' J_2')$$

$$(l^N\alpha J_1, l'^M\beta J_2; \gamma JJ| M_{D0}^{(3)} |l^N\alpha' J_1', l'^M\beta' J_2'; \gamma JJ)$$

where

$$(l^N \alpha J_1, l'^M \beta J_2; \gamma JJ | M_{D0}^{(3)} | l^N \alpha' J'_1, l'^M \beta' J'_2; \gamma JJ)$$

$$= \begin{pmatrix} J & 3 & J \\ -J & 0 & J \end{pmatrix} [J](-1)^{J+1}$$

$$\times \left[(-1)^{J_1 + J_2} \begin{Bmatrix} J & 3 & J \\ J'_1 & J_2 & J_1 \end{Bmatrix} \delta(J_2, J'_2) \delta(\beta, \beta') (l^N \alpha J_1 \| M_D^{(3)} \| l^N \alpha' J'_1) \right.$$

$$\left. + (-1)^{J'_1 + J'_2} \begin{Bmatrix} J & 3 & J \\ J'_2 & J_1 & J_2 \end{Bmatrix} \delta(J_1, J'_1) \delta(\alpha, \alpha') (l'^M \beta J_2 \| M_{D0}^{(3)} \| l'^M \beta' J'_2) \right].$$

(IX-15)

As before, when J_1 and J_2 are good quantum numbers, b and c can be expressed in terms of the b and c values for the different shells:

$$b = \frac{3}{(2J+3)(2J+2)} \left[\left(\frac{k_1(k_1+1) - \tfrac{4}{3}(J)(J+1)(J_1)(J_1+1)}{J_1(2J_1-1)} \right) b(l^N \phi J_1) \right.$$

$$\left. + \left(\frac{k_2(k_2+1) - \tfrac{4}{3}J(J+1)(J_2)(J_2+1)}{J_2(2J_2-1)} \right) b(l'^M \varepsilon J_2) \right], \quad \text{(IX-16)}$$

$$c = \frac{-5}{(J+1)(J+2)(2J+3)} \left[(k_1^3 + 4k_1^2 + \tfrac{4}{5}k_1\{-3(J_1)(J_1+1)(J)(J+1) \right.$$

$$+ J_1(J_1+1) + J(J+1) + 3\} - 5(J_1)(J_1+1)(J)(J+1))$$

$$\times [(2J_1)(2J_1-1)(2J_1-2)]^{-1} c(l^N \phi J_1)$$

$$+ (k_2^3 + 4k_2^2 + \tfrac{4}{5}k_2\{-3(J_2)(J_2+1)(J)(J+1) + J_2(J_2+1)$$

$$+ J(J+1) + 3\} - 5(J_2)(J_2+1)(J)(J+1))[(2J_2)(2J_2-1)(2J_2-2)]^{-1}$$

$$\left. \times c(l^N \phi J_2) \right]. \quad \text{(IX-17)}$$

The parameters k_1 and k_2 are given by

$$k_1 = J_2(J_2+1) - J(J+1) - J_1(J_1+1)$$

and

$$k_2 = J_1(J_1+1) - J(J+1) - J_2(J_2+1);$$

$b(l^N \phi J_1)$ by

$$b(l^N \phi J_1) = 2eQ \left[\frac{(2J_1)(2J_1-1)}{(2J_1+3)(2J_1+2)(2J_1+1)} \right]^{1/2} (l^N \phi J_1 \| Q_D^{(2)} \| l^N \phi J_1);$$

and $c(l^N \phi J_1)$ by

$$c(l^N \phi J_1) = -\Omega \left[\frac{(J_1)(2J_1-1)(J_1-1)}{(J_1+1)(J_1+2)(2J_1+3)(2J_1+1)} \right]^{1/2} (l^N \phi J_1 \| M_D^{(3)} \| l^N \phi J_1).$$

5. HYPERFINE STRUCTURE OF SAMARIUM

The $(4f)^6$ ^7F_J states of the ground term of Samarium have been extensively studied using a variety of techniques, both experimental and theoretical. The energy levels of the low-lying states were determined spectroscopically by Albertson;[180] Pichanick and Woodgate[181] studied the g_J values of these states. Conway and Wybourne[182] used these data to determine the wavefunctions for the low-lying states by means of parametric techniques such as those described in Section III-10. Hyperfine structure of the $J = 0$ to $J = 4$ were studied by Woodgate[100] using atomic beam techniques. This study was extended to the $J = 5$ state by Robertson, et al.[183] Finally, Rosen[97] has calculated relativistic radial integrals which aid in the analysis of the hyperfine structure. Because of the existence of all of this data, Samarium provides an interesting example to illustrate many of the results of the previous sections.

The analysis of the measured hyperfine structure should, in principal, begin by the fitting of observed energy levels (or, more accurately, the observed differences between energy levels) with the Hamiltonian of Eq. (VII-12). However, as noted in Section VII-4, one generally will not have available a computer program which can fit energy levels with such a complicated Hamiltonian. One must, instead, use programs which fit the energy levels with the Hamiltonian given by Eq. (VI-6). The first step in the analysis must be, therefore, to subtract from the observed transition frequencies the contribution from the second-order terms of Section VII-4 which are not simply absorbed into the first-order terms. All of the second-order terms which are not absorbed are, in fact, proportional to either **H** or $(\mathbf{HH})^{(2)}$, where **H** is an externally applied field. Thus, the first parameters which must be calculated are $\mathbf{T}^{(k)}$ (Eq. (VII-10)) and $\mathbf{Z}^{(2)}$ (Eq. (VII-11)). In order to calculate these parameters, the wavefunctions of the states are required. Table IX-8 gives the results obtained by Conway and Wybourne[182,183] from the analysis of the data of Albertson,[180] and Pichanick and Woodgate.[181] It should be noted, however, that these wavefunctions were obtained by assuming that the Slater integrals were related by their hydrogenic ratios, and ignoring relativistic effects such as spin-spin, etc. Thus, only F^2 and the spin-orbit constant were varied to obtain these results. One notes that L and S are very good quantum numbers for these states. Woodgate[100] evaluated $\mathbf{Z}^{(2)}$ using these wavefunctions, obtaining for the two lowest states

$$(^7F_1\|Z^{(2)}\|^7F_1) = -(0.9675) \times (0.747557)10^{-6}\mu_0^2$$
$$(^7F_2\|Z^{(2)}\|^7F_2) = -(0.9702) \times (0.286549)10^{-6}\mu_0^2$$

where the four decimal number represents the effect of the breakdown of L

Table IX-8. Intermediate Coupling Wavefunctions for Sm, $|f^6\ {}^7F_J\rangle = \sum_i \alpha_i(vWUSLJ)|f^6vWUSLJ\rangle$

v	W	U	SL	$J=0$	$J=1$	$J=2$	$J=3$	$J=4$	$J=5$
6	(100)	(10)	^7F	0.9712	0.9764	0.9833	−0.9884	−0.9904	0.9891
6	(210)	(11)	^5P	0	−0.0036	−0.0040	0.0034	0	0
6	(210)	(20)	^5D	0.0022	0.0054	0.0103	−0.0134	−0.0120	0
6	(210)	(21)	^5D	0.1582	0.1410	0.1116	−0.0772	−0.0418	0
4	(111)	(20)	^5D	−0.1734	−0.1579	−0.1308	0.0967	0.0575	0
6	(210)	(21)	^5F	0	0.0248	0.0445	−0.0615	−0.0724	0.0694
4	(111)	(10)	^5F	0	0.0156	0.0273	−0.0370	−0.0420	0.0378
6	(210)	(20)	^5G	0	0	−0.0047	0.0115	0.0222	−0.0390
6	(210)	(21)	^5G	0	0	−0.0136	0.0310	0.0535	−0.0814
4	(111)	(20)	^5G	0	0	0.0148	−0.0332	−0.0568	0.0852
6	(210)	(11)	^5H	0	0	0	−0.0015	−0.0036	0.0069
6	(210)	(21)	^5H	0	0	0	−0.0022	−0.0051	0.0088
6	(210)	(20)	^5I	0	0	0	0	0.0004	0
6	(221)	(31)	^3P	0.0281	0.0204	0.0097	0	0	0
4	(21:1)	(30)	^3P	0.0231	0.0164	0.0067	0	0	0
2	(110)	(11)	^3P	−0.0210	−0.0154	−0.0076	0	0	0
6	(221)	(31)	^3H	0	0	0	0	−0.0022	0
4	(211)	(21)	^3H	0	0	0	0	−0.0014	0

and S as good quantum numbers, that is, this number would be one if L and S were exactly good quantum numbers. The calculation of $\mathbf{T}^{(k)}$ is more complicated as it involves the exact form of the dipole and quadrupole interactions. The results of Section IV-4 and V-6 can be combined by writing

$$\mathbf{M}^{(1)} = \mu_0 \sum_i \left(\frac{4}{3} \left\langle \frac{\delta(r)}{r^2} \right\rangle_S \mathbf{s}_i + 2 \left\langle \frac{1}{r^3} \right\rangle_L \mathbf{l}_i - 2\sqrt{10} \left\langle \frac{1}{r^3} \right\rangle_{sC} (\mathbf{s}_i \mathbf{C}_i^{(2)})^{(1)} \right); \quad \text{(IX-18)}$$

the results of Sections IV-4 and V-8 can likewise be combined to yield

$$\mathbf{Q}^{(2)} = e \sum_i \left(\left\langle \frac{1}{r^3} \right\rangle_{sl} (\mathbf{s}_i \mathbf{l}_i)^{(2)} + \left\langle \frac{1}{r^3} \right\rangle_{Cl} (\mathbf{s}_i (\mathbf{C}_i^{(4)} \mathbf{l}_i)^{(3)})^{(2)} - \left\langle \frac{1}{r^3} \right\rangle_{Q} \mathbf{C}_i^{(2)} \right), \quad \text{(IX-19)}$$

where each of the radial parameters contains not only relativistic, but also core polarization effects.

The calculation of the contribution to $\mathbf{T}^{(k)}$ produced by the dipole interaction leads to an interesting and important result. Matrix elements of $\mathbf{L}$ off-diagonal in J are exactly the negative of those of $\mathbf{S}$ evaluated between the same off-diagonal states (Eq. (IX-4)). Thus the dipole contribution to $(J \| T^{(k)} \| J)$ is of the form

$$\sum_{J' \neq J} a_{JJ'} (J' \| \left[\frac{2}{3} \left\langle \frac{\delta(r)}{r^2} \right\rangle_S \mathbf{S} + \left\langle \frac{1}{r^3} \right\rangle_L \mathbf{L} - \sqrt{10} \left\langle \frac{1}{r^3} \right\rangle_{sC} \sum_i (\mathbf{s}_i \mathbf{C}_i^{(2)})^{(1)} \right] \| J)$$

$$= b_J \left(\frac{2}{3} \left\langle \frac{\delta(r)}{r^2} \right\rangle_S - \left\langle \frac{1}{r^3} \right\rangle_L \right) + c_J \left\langle \frac{1}{r^3} \right\rangle_{sC}$$

that is, always involves the difference between $\frac{2}{3} \langle \delta(r)/r^2 \rangle_S$ and $\langle 1/r^3 \rangle_L$. The calculation of the quadrupole contribution to $\mathbf{T}^{(k)}$ is much simpler, because the terms $\langle 1/r^3 \rangle_{sl}$ and $\langle 1/r^3 \rangle_{Cl}$ are almost certainly too small to contribute to second-order effects. The results of Woodgate's calculations of the dipole- and quadrupole-field perturbation are summarized in Table IX-9.

At low fields, the second-order terms discussed above will produce only small effects. Thus, one can assume

$$\left\langle \frac{1}{r^3} \right\rangle_{sC} = \left\langle \frac{1}{r^3} \right\rangle_L = \left\langle \frac{1}{r^3} \right\rangle_Q = \left\langle \frac{1}{r^3} \right\rangle, \quad \left\langle \frac{\delta(r)}{r^2} \right\rangle_S = 0.$$

The total dipole-field term (Eq. (VII-10)), then is simply proportional to $\langle 1/r^3 \rangle \mu_I$, a quantity which can be roughly evaluated using Eq. (IV-37)

$$A_J = \frac{2\mu_I \mu_0}{I} \left\langle \frac{1}{r^3} \right\rangle (^7F_J \| L - \sqrt{10} \sum_i (\mathbf{s}_i \mathbf{C}_i^{(2)})^{(1)} \|^7 F_J) [J(J+1)(2J+1)]^{-\frac{1}{2}}$$

and the measured uncorrected A_J's: one finds (Sm^{147})

$$\frac{2\mu_I \mu_0}{I} \left\langle \frac{1}{r^3} \right\rangle \approx -140 \text{ mc/sec.}$$

Table IX-9. Reduced Matrix Elements of $T^{(k)} \times 10^6$

J	k	$T^{(k)} \times 10^6$
1	0	$-(1.0553)(0.014866)2\mu_0^2 \left(\left\langle \frac{1}{r^3} \right\rangle_L - \frac{2}{3}\left\langle \frac{\delta(r)}{r^2} \right\rangle_S \right)$
		$-(2.1283)(0.006808)2\mu_0^2 \left\langle \frac{1}{r^3} \right\rangle_{SC}$
	1	$+(0.9710)(0.037376)\mu_0 e \left\langle \frac{1}{r^3} \right\rangle_Q$
	2	$-(0.9675)(1.179253)2\mu_0^2 \left(\left\langle \frac{1}{r^3} \right\rangle_L - \frac{2}{3}\left\langle \frac{\delta(r)}{r^2} \right\rangle_S \right)$
		$+(1.3365)(0.201185)2\mu_0^2 \left\langle \frac{1}{r^3} \right\rangle_{SC}$
2	0	$-(1.2308)(0.004716)2\mu_0^2 \left(\left\langle \frac{1}{r^3} \right\rangle_L - \frac{2}{3}\left\langle \frac{\delta(r)}{r^2} \right\rangle_S \right)$
		$-(1.6272)(0.011621)2\mu_0^2 \left\langle \frac{1}{r^3} \right\rangle_{SC}$
	1	$+(0.9827)(0.028832)\mu_0 e \left\langle \frac{1}{r^3} \right\rangle_Q$
	2	$-(0.9704)(0.451938)2\mu_0^2 \left(\left\langle \frac{1}{r^3} \right\rangle_L - \frac{2}{3}\left\langle \frac{\delta(r)}{r^2} \right\rangle_S \right)$
		$+(1.2381)(0.070611)2\mu_0^2 \left\langle \frac{1}{r^3} \right\rangle_{SC}$
	3	$+(0.9679)(0.045090)\mu_0 e \left\langle \frac{1}{r^3} \right\rangle_Q$

The total quadrupole-field correction is, in this approximation, proportional to $\langle 1/r^3 \rangle_Q$. From Eq. (IV-40) and the uncorrected b_J's, we find (Sm^{147})

$$-\tfrac{1}{2} e^2 Q \left\langle \frac{1}{r^3} \right\rangle_Q \approx 150 \text{ mc/sec.}$$

Using these parameters, the shifts in transition frequency can be calculated at low field. The measured frequencies are then corrected, and the corrected values fit with the Hamiltonian of Eq. (VI-6). The results of such a calculation by Woodgate[100] for the $J=1$ to $J=4$ states of Samarium are shown in Table IX-10.

Table IX-10. Observed Hyperfine Interaction Constants

	Sm^{147}		Sm^{149}	
J	A	b	A	b
1	−33.4972(1)	−58.7544(8)	−27.6139(1)	16.9196(4)
2	−41.1864(2)	−62.2576(32)	−33.9520(2)	17.9628(32)
3	−50.2410(1)	−33.7032(16)	−41.4184(4)	9.7296(64)
4	−59.7075(20)	21.2148(720)	−49.2182(28)	−6.1764(800)
5	−69.136(1)	100.560(60)	−56.992(2)	−29.048(92)

The Eqs. (VII-9) can now be used in order to determine the "real" hyperfine interaction constants. These calculations are straightforward, but involve the unknown radial parameters $\langle \delta(r)/r^2 \rangle_S$, $\langle 1/r^3 \rangle_L$, $\langle 1/r^3 \rangle_{sL}$, and $\langle 1/r^3 \rangle_Q$. A first approximation for these second-order effects can be obtained by employing the same approximations used for the field-dependent terms. The resulting correct A and b values as obtained by Woodgate[100] for the $J = 1$ to 4 states are given in Table IX-11.

Table IX-11. Corrected Hyperfine Interaction Constants

	Sm^{147}		Sm^{149}	
J	A	b	A	b
		a. Woodgate[100]		
1	−33.4935(1)	−58.6920(8)	−27.6108(1)	16.9624(4)
2	−41.1844(2)	−62.2260(32)	−33.9508(2)	17.9872(32)
3	−50.2401(1)	−33.6812(16)	−41.4176(4)	9.7488(64)
4	−59.7068(20)	21.2304(720)	−49.2177(28)	−6.1612(800)
		b. Robertson et al.[183]		
1	−33.4931(1)	−58.6848(8)	−27.6106(1)	16.9668(4)
2	−41.1839(7)	−62.2184(56)	−33.9508(3)	17.9884(32)
3	−50.2396(4)	−33.6776(36)	−41.4176(4)	9.7480(64)
4	−59.7067(20)	21.2336(72)	−49.2176(28)	−6.1936(800)
5	−69.1354(13)	100.5792(52)	−56.9922(22)	−29.0312(920)

One can now fit these real A and b values with Eqs. (IX-18), (IX-19), leaving the radial integrals as variable parameters. A problem arises, however, because we are dealing with states which are almost pure $|^7F_J\rangle$, that is, for which $L = S$. Then, as we see using Eq. (IX-3)

$$(^7F_J\|\sum_i \left(\left\langle \frac{1}{r^3}\right\rangle_L l_i + \frac{2}{3}\left\langle \frac{\delta(r)}{r^2}\right\rangle_S s_i\right)\|^7F_J)$$

$$= \tfrac{1}{2}[J(J+1)(2J+1)]^{1/2}\left(\left\langle \frac{1}{r^3}\right\rangle_L + \frac{2}{3}\left\langle \frac{\delta(r)}{r^2}\right\rangle_S\right).$$

That is, the same linear combination of $\langle 1/r^3\rangle_L$ and $\langle \delta(r)/r^2\rangle_S$ appears in the expression for A_J irrespective of J. The exact same linear combination will not appear in the actual calculation, of course, because the $|^7F_J\rangle$ are not completely pure: as shown by Eq. (IX-3), the contribution to the matrix element of L due to breakdown of L and S will be exactly the negative of the contribution to the matrix element of S from the same source. Nevertheless, this breakdown of LS coupling produces an effect of less than 1% in the coefficients multiplying $\langle 1/r^3\rangle_L$ and $\langle \delta(r)/r^2\rangle_S$. It is thus impossible to determine these parameters individually with any accuracy by fitting the A_J's; one must solve instead for the combination $(\langle 1/r^3\rangle_L + \tfrac{2}{3}\langle \delta(r)/r^2\rangle_S)$. Woodgate carried out such a calculation, as well as a similar fitting of the b_J's assuming that $\langle 1/r^3\rangle_{sl}$ and $\langle 1/r^3\rangle_{cl}$ are zero.

In addition to the low-field measurements already mentioned, Woodgate performed a triple-resonance experiment at high field in order to directly determine the nuclear dipole moments of Sm^{147} and Sm^{149}. In this experiment, a transition of the type $M_I \leftrightarrow M_I - 1$, $\Delta M_J = 0$ is observed (high field notation). Using the results of Section VI-1, we see that the frequency of such a transition is approximately

$$h\nu = AM_J + 3b\frac{[3M_J^2 - J(J+1)]}{4IJ(2I-1)(2J-1)}(2M_I - 1) - g_I\mu_0 H$$

where A and b are the measured, not the "real," dipole and quadrupole constants. This equation is, of course, not exact, since it is valid only if M_I and M_J are rigorously good quantum numbers; it does, however, demonstrate the variables involved. In addition, the second-order field dependent perturbations will contribute to the observed frequency. Thus the actual observed frequency will be of the form

$$h\nu = \alpha + g_I\beta + g_I\left(\frac{2}{3}\left\langle\frac{\delta(r)}{r^2}\right\rangle_S - \left\langle\frac{1}{r^3}\right\rangle_L\right)\gamma$$

where α, β, and γ are functions of A, b, and H; functions which can be obtained from the results of the previous calculation. Unfortunately, although carried out in both the $J=1$ and $J=2$ states of Sm^{147}, this measurement could only be carried out in the $J=1$ state of Sm^{149}. The difficulty arises because there are three unknowns in the above equation for the transition frequency: g_I, $\langle \delta(r)/r^2 \rangle_S$ and $\langle 1/r^3 \rangle_L$. The parameter $\langle 1/r^3 \rangle_L$ can be considered constant for the two isotopes, but both g_I and $\langle \delta(r)/r^2 \rangle_S$ (see Section VII-2) can vary from isotope to isotope. In order to interpret this data, therefore, one must make some simplifying assumptions. Let us assume that $\langle \delta(r)/r^2 \rangle_S$ is the same for both isotopes, and that $g_I(147)/g_I(149) = A_1(147)/A_1(149)$, that is, that there is no hyperfine anomaly. We will later consider the validity of this assumption. With these assumptions, Woodgate was able to determine $g_I(147) = -125 \times 10^{-6}$ and $2\mu_0(\langle 1/r^3 \rangle_L - \tfrac{2}{3} \langle \delta(r)/r^2 \rangle_S) = 1.15 \times 10^6$ mc/sec. These results in combination with those obtained from the low-field experiments, enabled Woodgate to ascertain the parameters shown in Table IX-12.

Table IX-12. Parameters Obtained from Experimental Data

	Woodgate[100]	Robertson et al.[183]
$\langle \frac{1}{r^3} \rangle_L$	$6.390(6) a_0^{-3}$	$6.382(7) a_0^{-3}$
$\langle \frac{1}{r^3} \rangle_{sc}$	$6.513(12) a_0^{-3}$	$6.461(11) a_0^{-3}$
$\langle \frac{\delta(r)}{r^2} \rangle_s$	$-0.312(9) a_0^{-3}$	$-0.325(7) a_0^{-3}$
$\frac{e^2 Q(147)}{2h} \langle \frac{1}{r^3} \rangle_{sl}$		0.271(6) Mc/sec
$\frac{e^2 Q(147)}{2h} \langle \frac{1}{r^3} \rangle_{cl}$		2.59(3) Mc/sec
$\frac{e^2 Q(147)}{2h} \langle \frac{1}{r^3} \rangle_Q$		$-151.73(3)$ Mc/sec
$\mu_I(147)$	$-0.8074(7)$ nm	

The process above is clearly not exact, since at each step one is required to make certain simplifying assumptions. At this point, therefore, the values we have obtained above should be used to improve the original assumptions. The calculations can then be repeated until the process converges. For example, the second order corrections of Eqs. (VII-9) can be redetermined using the value of $(\langle 1/r^3 \rangle_L - \tfrac{2}{3} \langle \delta(r)/r^3 \rangle_S)$ determined in the last step above and the values of $\langle 1/r^3 \rangle_{sc}$ and $\langle 1/r^3 \rangle_Q$ obtained in the next-to-last step. New values of the corrected A and b can then be determined. This step has been carried out by Robertson et al.;[183] the new values for A and b are shown in Table IX-11. (The uncertainties shown in this table allow for a second order octupole-like term in the Hamiltonian.) In addition, Robertson, et al. measured the hyperfine structure in the $J = 5$ state. The corrected results of this measurement are also shown in this table. Robertson, et al. next determined new values of the radial parameters by fitting these new corrected A values with the interaction of Eq. (IX-18), while simultaneously fitting the equation obtained by Woodgate from his high field data, $2\mu_0(\langle 1/r^3\rangle_L - \frac{2}{3}\langle \delta(r)/r^2\rangle_S) = 1.15 \times 10^6$ mc/sec. The radial parameters obtained in this analysis are shown in Table IX-12. Robertson, et al. also fit the corrected b values with the interaction of Eq. (IX-19) in order to obtain values for all three quadrupole radial integrals. These results are also shown in this table.

In principle, these new numbers should now be used to re-analyze Woodgate's high field results. However, Woodgate noted that the second-order corrections involving $\langle 1/r^3\rangle_{SC}$ and $\langle 1/r^3\rangle_Q$ were very small. Thus repeating that portion of the analysis would probably not lead to any significant change in the relationship between $\langle 1/r^3\rangle_L$ and $\langle \delta(r)/r^2\rangle_S$. Before leaving this part of the calculation, however, let us note that, because of difficulties in fitting the data, Robertson et al. concluded that the wavefunctions of Wybourne and Conway contained errors in the third decimal place.

Table IX-13. Radial Parameters Calculated by Rosen[97]

	HFS	OHFS
$\langle \frac{1}{r^3} \rangle_{01}$	6.995	6.225
$\langle \frac{1}{r^3} \rangle_{12}$	7.434	6.709
$\langle \frac{\delta(r)}{r^2} \rangle_{10}$	−0.339	−0.345
$\langle \frac{1}{r^3} \rangle_{11}$	−0.015	−0.016
$\langle \frac{1}{r^3} \rangle_{13}$	−0.141	−0.138
$\langle \frac{1}{r^3} \rangle_{02}$	6.973	6.238

Rosen[97] has used a Hartree-Fock-Slater procedure to calculate the relativistic radial integrals for Sm. These numbers are shown in Table IX-13. The values labeled HFS were obtained using a Slater exchange term; those labeled OHFS were obtained using a parametrized exchange term which was chosen to minimize the energy. Agreement between calculated and measured values is clearly not perfect for either type of potential. Perhaps the most interesting result is the close agreement between the two calculated $\langle \delta(r)/r^2 \rangle_S$ values and the measured value. This probably indicates that there is very little s electron core polarization present in Sm, and supports the assumption that no hyperfine anomaly is present. On the other hand, the ratios

$$\left\langle \frac{1}{r^3} \right\rangle_L \bigg/ \left\langle \frac{1}{r^3} \right\rangle_{sC}$$

calculated with both potentials are roughly equal, and much larger than the experimental ratio. It is probable, therefore, that configuration interaction is affecting one or both of these values.

Appendix. Nonrelativistic Limits

In taking non-relativistic limits, we have always made the assumption that

$$E + mc^2 - U(r) + \frac{Ze^2}{r} = 2mc^2 + O\left(\frac{v}{c}\right)^2.$$

By doing so, we have been able to replace the radial wavefunction of the small component, G_{nlj}, with an expression in F_{nlj}, the radial wavefunction of the large component:

$$G_{nlj} = -\frac{\mu_0}{e}\left(\frac{d}{dr} - \frac{\kappa}{r}\right)F_{nlj}. \tag{III-13}$$

Clearly, however, this equation can be valid to order $(v/c)^2$ only if $-U(r) + Ze^2/r \ll mc^2$. This inequality should never be violated by $U(r)$, which has no singular points, but will be violated by Ze^2/r at small r. Thus, we must conclude that Eq. (III-13) is not valid for distances of the order of $2r_0 = Ze^2/mc^2 = Z2.8\ 10^{-13}$ cm. This distance is very small compared to the classical distance from the nucleus of an atomic electron ($\sim 10^{-8}$ cm.), and one is normally justified in ignoring any errors contained in (III-13). However, in Section IV-6, we use Eq. (III-13) to determine effects caused by a non-relativistic electron at $r = 0$; some discussion would, therefore, seem necessary in order to ascertain the error introduced into our results through (possibly) indiscriminate use of (III-13).

The origin of the difficulty can be easily seen if we write the usual first order Dirac equation for an electron in a central field (Eq. (III-8)) in terms of a second order equation[184]

$$\left[E_0 + \frac{\hbar^2}{2m}\nabla^2 + e\phi + \frac{1}{2mc^2}(E_0 + e\phi)^2 + \frac{ie\hbar}{mc}(\mathbf{A}\cdot\mathbf{V})\right.$$
$$\left. - \frac{e^2}{2mc^2}A^2 - \frac{e\hbar}{2mc}(\boldsymbol{\sigma}\cdot\mathbf{H}) + \frac{ie\hbar}{2mc}(\boldsymbol{\alpha}\cdot\mathbf{E})\right]\psi = 0.$$

Appendix. Nonrelativistic Limits

The only term in this equation which couples the large and small components is $\boldsymbol{\alpha} \cdot \mathbf{E}$. When ϕ (and $\mathbf{E} = -\nabla\phi$) is small, one can replace this term by

$$\frac{\mu_0}{2mc}(i\mathbf{E} \cdot \mathbf{p} - \boldsymbol{\sigma} \cdot \mathbf{E} \times \mathbf{p})$$

with the resulting equation being one in the large component only. Having done this, the usual Schrodinger for an atomic electron in a region having no external fields is obtained by putting ϕ equal to the central field potential and considering only the first three terms above. Clearly, in such a case one is neglecting terms of order ϕ/mc^2. Thus the Schrodinger equation itself cannot be considered to be correct in regions of large potential ($\phi \sim mc^2$) even if the energy of the electron is non-relativistic. Therefore one difficulty which must be expected to arise in taking limits commensurate with those involved in the Schrodinger equation is that the resulting equations will be invalid in regions of large potential fields. Our limiting procedure clearly falls into this category.

Having shown that our difficulties are not unusual, it is nonetheless interesting to estimate the errors involved in the procedure we have followed. Let us, therefore, consider the taking of the non-relativistic limit for the dipole integral (Section IV-6). Here, we are interested in the integral

$$\int \frac{F_{nlj} G_{nlj'}}{r^2} dr = -\hbar c \int \frac{F_{nlj}}{r^2} \left(E + mc^2 - U(r) + \frac{Ze^2}{r} \right)^{-1} \left(\frac{d}{dr} - \frac{\kappa'}{r} \right) F_{nlj'} \, dr \quad \text{(A-1)}$$

where the expression on the right above results from use of Eq. (III-11). Let us now make the approximation

$$E + mc^2 - U(r) + \frac{Ze^2}{r} = 2mc^2 + \frac{Ze^2}{r} + O\left(\frac{v}{c}\right)^2 \quad \text{(A-2)}$$

which is the approximation made above, retaining, however, the singular potential term. Then, proceeding as before, we write (A-1) as

$$-\hbar c \int \frac{F_{nlj}}{2mc^2 r^2 + Ze^2 r} \left(\frac{d}{dr} - \frac{\kappa'}{r} \right) F_{nlj'} \, dr$$

$$= -\frac{\hbar}{2mc} \left[\int \frac{F_{nlj}}{(r^2 + rr_0)^{1/2}} \frac{d}{dr} \frac{F_{nlj'}}{(r^2 + rr_0)^{1/2}} \, dr \right.$$

$$\left. + \frac{1}{2} \int \frac{F_{nlj} F_{nlj'}(2r + r_0)}{(r^2 + rr_0)^2} \, dr - \kappa' \int \frac{F_{nlj} F_{nlj'}}{r(r^2 + rr_0)} \, dr \right]. \quad \text{(A-3)}$$

In order to evaluate these terms, we must determine the equation satisfied by F_{nlj} in this limit. Proceeding as before, one finds

Appendix. Nonrelativistic Limits

$$\left[-\frac{\hbar^2}{2m}\left(\frac{d^2}{dr^2} - \frac{l(l+1)}{r^2}\right) - E_0 - \frac{Ze^2}{r} + U(r)\right]F_{nlj}$$

$$= \left[\frac{1}{2mc^2}\left(+E_0 + \frac{Ze^2}{r} - U(r)\right)\left(\frac{Ze^2}{r}\right) + \frac{\hbar^2 r_0}{2m(r+r_0)r}\left(\frac{d}{dr} - \frac{\kappa}{r}\right)\right]F_{nlj}.$$

Bethe and Salpeter[184] have considered an equation such as this, reaching the conclusion that F_{nlj} differs in this limit from R_{nl} by terms of order $(Z\alpha)^2$, which is of order $(v/c)^2$. Thus, introducing errors of order $(Z\alpha)^2$, we can replace F_{nlj} by R_{nl}. Equation (A-3) then becomes

$$-\frac{\mu_0}{e}\frac{R_{nl}^2}{2(r^2+rr_0)}\bigg|_{r=0} + \frac{\mu_0}{e}(\kappa-1)\int\frac{rR_{nl}^2}{(r^2+rr_0)^2}\,dr$$

$$+ \frac{\mu_0}{e}\left(\kappa - \frac{1}{2}\right)\int\frac{r_0 R_{nl}^2}{(r^2+rr_0)^2}\,dr. \quad (A-4)$$

The first term in (A-4) is, of course, identical to

$$-\frac{\mu_0}{2e}\left|\frac{R_{nl}}{r}\right|^2\bigg|_{r=0}.$$

One can estimate the error introduced by replacing the remaining terms in (A-4) by the expressions of Section IV-6 by assuming that, in the region in which differences occur, the wavefunction is essentially hydrogenic. The integrals involved can then easily be carried out explicitly. The results indicate that an error of order $(Z\alpha)^2 \ln(Z\alpha)$ results from replacing the two integrals of (A-4) by the integral of Section IV-6,

$$\frac{\mu_0}{e}(\kappa - 1)\int\frac{R_{nl}^2}{r^3}\,dr.$$

We have, then, the result that by ignoring the singularity of the potential, we have introduced errors of the order $(Z\alpha)^2(\ln(Z\alpha) + C)$, C of order 1, into our results. However, we have already chosen to ignore terms of this order in (A-3). Thus, we find that use of Eq. (III-13) does not lead to errors larger than $(Z\alpha)^2$, even though the singularity of the potential is not correctly taken into account.

Finally, we note that using the substitution III-13 in the Breit equation insures that we will obtain only those nonrelativistic terms which refer to two positive energy particles. This prevents the appearance of spurious terms which arise when the Breit interaction is taken to its nonrelativistic limit for negative energy particles.[47, 184]

References

1. A. Michelson, *Phil. Mag.* **31**, 338 (1891).
2. C. Fabry and A. Perot, *Ann. Chim. et Phys.* **12**, 459 (1897).
3. O. Lummer and E. Gehrcke, *Ann. Phys.* **10**, 457 (1903).
4. W. Pauli, *Naturwissenschaften* **12**, 74 (1924).
5. E. Back and S. Goudsmit, *Z. Physik* **43**, 321 (1927); **47**, 174 (1928).
6. D. A. Jackson, *Proc. Roy. Soc.* (*London*) **A121**, 432 (1928).
7. De Broglie, *Phil. Mag.* **47**, 446 (1924).
8. E. Schrodinger, *Ann. Physik* **79**, 361, 489, 734 (1926); **80**, 437 (1926); **81**, 109 (1926).
9. E. Fermi, *Z. Physik* **60**, 320 (1930).
10. G. Breit, *Phys. Rev.* **35**, 1447 (1930); **37**, 51 (1931).
11. S. Goudsmit, *Phys. Rev.* **37**, 663 (1931).
12. G. Breit, *Phys. Rev.* **38**, 463 (1931).
13. G. Racah, *Z. Physik* **71**, 431 (1931); *Nuevo Cimento* **8**, 178 (1931).
14. H. Schuler and T. Schmidt, *Z. Physik* **94**, 457 (1935).
15. H. B. G. Casimir, *On the Interaction between Atomic Nuclei and Electrons* (Teyler's Tweede Genootschap, Haarlem, 1936; W. H. Freeman, San Francisco, 1963).
16. H. B. G. Casimir and G. Karreman, *Physica* **9**, 494 (1942).
17. H. Schuler and J. Keyston, *Naturwissenschaften* **19**, 320 (1931).
18. H. Kopfermann, *Naturwissenschaften* **19**, 400 (1931).
19. G. Racah, *Phys. Rev.* **62**, 438 (1942).
20. G. Racah, *Phys. Rev.* **63**, 367 (1943).
21. G. Racah, *Phys. Rev.* **76**, 1352 (1949).
22. R. E. Trees, *Phys. Rev.* **92**, 308 (1953).
23. C. Schwartz, *Phys. Rev.* **97**, 380 (1955).
24. P. G. H. Sandars and J. Beck, *Proc. Roy. Soc.* (*London*) **A289**, 97 (1955).
25. I. Rabi, S. Millman, P. Kusch, and J. R. Zacharias, *Phys. Rev.* **55**, 526 (1939).
26. N. F. Ramsey, *Molecular Beams* (Oxford University Press, London, 1956).
27. J. Brossel and A. Kastler, *Compt. Rend.* **229**, 1213 (1949). R. A. Bernheim, *Optical Pumping: An Introduction* (W. A. Benjamin, Inc., New York, 1965). H. R. Schlossberg and A. Javan, *Phys. Rev. Lett.* **17**, 1242 (1966).

28. F. Block, W. W. Hansen, and M. Packard, *Phys. Rev.* **69**, 37 (1946). E. M. Purcell, H. Torrey, and R. V. Pound, *Phys. Rev.* **69**, 37 (1946). C. P. Schlichter, *Principles of Magnetic Resonance* (Harper and Row, New York, 1963). A. Abragam, *The Principles of Nuclear Magnetism* (Oxford University Press, London, 1961).
29. E. J. Zavoiski, *J. Phys. U.S.S.R.* **9**, 211 (1945). W. Low in *Solid State Physics*, Suppl. 2 (1960). W. Low (ed.), *Paramagnetic Resonance*, Vols. I and II (Academic Press, New York, 1963).
30. R. L. Mössbauer, *Z. Physik* **151**, 124 (1958). G. K. Wertheim, *Mössbauer Effect: Principles and Applications* (Academic Press, New York, 1964). H. Frauenfelder, *The Mössbauer Effect* (W. A. Benjamin, Inc., New York, 1962).
31. S. G. Goren in *Hyperfine Interactions*, A. J. Freeman and R. B. Frankel, eds. (Academic Press, New York, 1967), p. 553. E. Matthias, *ibid.* p. 595.
32. E. U. Condon and G. H. Shortley, *The Theory of Atomic Spectra* (Cambridge University Press, London, 1963). J. C. Slater, *Quantum Theory of Atomic Structure*, Vols. I and II (McGraw Hill Book Company, Inc., New York, 1960).
33. A. R. Edmonds, *Angular Momentum in Quantum Mechanics* (Princeton University Press, Princeton, 1957).
34. J. D. Jackson, *Classical Electrodynamics* (John Wiley and Sons, Inc., New York, 1962).
35. M. A. Preston, *Physics of the Nucleus* (Addison-Wesley Publishing Company, Inc., Reading, Mass., 1962).
36. A. de-Shalit and I. Talmi, *Nuclear Shell Theory* (Academic Press, New York, 1963).
37. A. Bohr and B. R. Mottelson, *Nuclear Structure*, Vol. I (W. A. Benjamin, Inc., New York, 1969).
38. D. M. Brink and G. R. Satchler, *Angular Momentum*, 2nd edition (Clarendon Press, Oxford, 1968).
39. M. Rotenberg, R. Bivens, N. Metropolis, and J. K. Wooten, Jr., *The 3-j and 6-j Symbols* (The Technology Press, MIT, Cambridge, Mass., 1958).
40. B. R. Judd, *Operator Techniques in Atomic Spectroscopy* (McGraw-Hill Book Company, New York, 1963).
41. A. P. Yutsis, I. B. Levinson, and V. V. Vanagas, *Mathematical Apparatus of the Theory of Angular Momentum* (Israel Program for Scientific Translations, Jerusalem, 1962).
42. L. C. Biedenharn, *J. Math. Phys.* **31**, 287 (1953). J. P. Elliot, *Proc. Roy. Soc. (London)* **A218**, 345 (1953).
43. B. R. Judd, *Second Quantization and Atomic Spectroscopy* (Johns Hopkins Press, Baltimore, 1967).
44. R. P. Feynman, *Phys. Rev.* **76**, 749 (1949); **76**, 769 (1949).
45. B. G. Wybourne, *Spectroscopic Properties of Rare Earths* (Interscience Publishers, New York, 1965).
46. A. I. Akhiezer and V. B. Berestetskii, *Quantum Electrodynamics* (Interscience Publishers, London, 1965). Translated by G. M. Volkoff.
47. G. Breit, *Phys. Rev.* **34**, 553 (1929).
48. P. A. M. Dirac, *Proc. Roy. Soc. (London)* **A117**, 610; **A118**, 351 (1928).
49. W. Pauli, *Z. Physik* **31**, 765 (1925).
50. R. D. Lawson and M. H. Macfarlane, *Nucl. Phys.* **66**, 80 (1965).

References

51. E. P. Wigner, *Group Theory* (Academic Press, New York, 1959). Translated by J. J. Griffin.
52. M. Hamermesh, *Group Theory and Its Application to Physical Problems* (Addison-Wesley Publishing Company, Inc., Reading, Mass., 1962).
53. G. Racah, *Group Theory and Spectroscopy*, in Springer Tracts in Modern Physics **37** (Springer-Verlag, Berlin, 1965), p. 28.
54. S. Lie and G. Scheffers, *Vorlesungen über kontinuierliche Gruppen* (Teubner Verlagsgesellschaft, Leipzig, 1893).
55. H. Weyl, *Math. Z.* **23**, 271 (1925); **24**, 328, 377 (1925).
56. E. Cartan, *Sur la Structure des Groupes de Transformations Finis et Continus* (thesis, Paris, 1894).
57. B. L. Van der Waerden, *Math. Z.* **37**, 446 (1933).
58. B. R. Judd, *Phys. Rev.* **162**, 28 (1967).
59. L. Armstrong, Jr., and B. R. Judd, *Proc. Roy. Soc.* (*London*) **A315**, 27, 39 (1970).
60. S. Feneuille, *J. Phys. Radium* **28**, 61 (1967).
61. L. Armstrong, Jr., *J. Math. Phys.* **7**, 1891 (1966); **9**, 1083 (1968).
62. P. B. Nutter, Raytheon Technical Memorandum No. T-544, 1964 (unpublished).
63. L. Armstrong, Jr., and S. Feneuille, *Phys. Rev.* **173**, 58 (1968).
64. H. H. Marvin, *Phys. Rev.* **71**, 102 (1947).
65. K. Rajnak and B. G. Wybourne, *Phys. Rev.* **132**, 280 (1963).
66. A. Messiah, *Quantum Mechanics*, Vols. I and II (North-Holland Publishing Company, Amsterdam, 1962). Translated by J. Potter.
67. L. Armstrong, Jr., and R. Marrus, *Phys. Rev.* **138**, B310 (1965).
68. F. R. Innes and C. W. Ufford, *Phys. Rev.* **111**, 194 (1958).
69. M. Blume and R. E. Watson, *Proc. Roy. Soc.* **A270**, 127 (1962); **A271**, 565 (1963).
70. B. R. Judd, *Proc. Phys. Soc.* **82**, 874 (1963).
71. K. Rajnak and B. G. Wybourne, *Phys. Rev.* **134**, A596 (1964).
72. S. Feneuille, *J. Phys.* (*Paris*) **28**, 497 (1967).
73. L. Armstrong, Jr., *J. Chem. Phys.* **51**, 129 (1969).
74. J. Bauche and B. R. Judd, *Proc. Phys. Soc.* **82**, 874 (1963).
75. B. R. Judd, in *La Structure Hyperfine Magnetique des Atoms et des Molecules* (Centre National de la Recherche Scientifique, Paris, 1967), p. 311.
76. F. Herman and S. Skillman, *Atomic Structure Calculations* (Prentice-Hall, Englewood Cliffs, N.J., 1963).
77. Y. Bordarier, B. R. Judd, and M. Klapisch, *Proc. Roy. Soc.* **A289**, 81 (1965).
78. R. K. Nesbet, *Phys. Rev.* **118**, 681 (1960).
79. N. Bessis, H. Lefebvre-Brion, and C. M. Moser, *Phys. Rev.* **124**, 1124 (1961).
80. H. F. Schaefer, III, R. A. Klemm, and F. E. Harris, *Phys. Rev.* **176**, 49 (1968); **181**, 137 (1969). H. F. Schaefer, III and R. A. Klemm, *Phys. Rev.* **188**, 152 (1969).
80a. P. S. Bagus, Bowen Liu, and H. F. Schaefer, III, *Phys. Rev. A* **2**, 555 (1970).
81. N. Bessis, H. Lefebvre-Brion, and C. M. Moser, *Phys. Rev.* **128**, 213 (1962); N. Bessis, H. Lefebvre-Brion, C. M. Moser, A. J. Freeman, R. K. Nesbet, and R. E. Watson, *Phys. Rev.* **135**, A588 (1964).

82. C. M. Moser, in *Hyperfine Interactions*, A. J. Freeman and R. B. Frankel, eds., (Academic Press, New York, 1967).
83. N. Bessis, H. Lefebvre-Brion, and C. M. Moser, *Phys. Rev.* **130**, 1441 (1963); N. Bessis, J. Picart, and J. P. Desclaux, *Phys. Rev.* **187**, 88 (1969).
84. A. J. Freeman and R. E. Watson in *Magnetism*, G. T. Rado and H. Suhl, eds. (Academic Press, New York, 1965). Vol. IIA, p. 167.
85. R. E. Watson and A. J. Freeman in *Hyperfine Interactions*, A. J. Freeman and R. B. Frankel, eds. (Academic Press, New York, 1967).
86. A. J. Freeman, P. Bagus, and R. E. Watson, in *La Structure Hyperfine Magnetique des Atoms et des Molecules* (Centre National de la Recherche Scientifique, Paris, 1967), p. 293.
87. J. Kerwin and E. A. Burke, *Phys. Rev.* **159**, 57 (1967).
88. P. O. Lowdin, *Phys. Rev.* **97**, 1474 (1955).
89. W. A. Goddard III, *Phys. Rev.* **157**, 81, 93 (1967); **176**, 106 (1968); **182**, 48 (1969).
90. R. M. Sternheimer, *Phys. Rev.* **86**, 316 (1952).
91. G. D. Gaspari, Wei-Mei Shyu, and T. P. Das, *Phys. Rev.* **134**, A852 (1964).
92. A. Abragam, J. Horowitz, and M. H. L. Pryce, *Proc. Roy. Soc.* **A230**, 169 (1955).
93. W. J. Childs, *Phys. Rev.* **160**, 9 (1967).
94. L. Armstrong, Jr., *Bull. Amer. Phys. Soc.* **10**, 1214 (1965).
95. W. Burton Lewis, J. B. Mann, D. A. Liberman, and D. T. Cromer, *J. Chem. Phys.* **53**, 809 (1970).
96. M. B. Bleaney, *Proceedings of the 3rd International Conference on Quantum Electronics* (Columbia University Press, New York, 1964), p. 595.
97. A. Rosen, *J. Phys.* **B2**, 1257 (1969).
98. W. J. Childs and L. S. Goodman, *Phys. Rev.* **156**, 64 (1967); **170**, 50 (1968). W. J. Childs, *Phys. Rev.* **156**, 71 (1967).
99. H. E. Radford and K. M. Evenson, *Phys. Rev.* **168**, 70 (1967).
100. G. K. Woodgate, *Proc. Roy. Soc.* **A293**, 117 (1966).
101. C. Bauche-Arnoult, *Proc. Roy. Soc.* **A322**, 361 (1971).
102. J. S. M. Harvey, *Proc. Roy. Soc.* **A285**, 581 (1965).
103. H. P. Kelly, *Phys. Rev.* **173**, 142 (1968); **180**, 55 (1969).
104. H. Wolter, *Z. Phys.* **205**, 492 (1967).
105. R. M. Sternheimer, *Phys. Rev.* **80**, 102 (1950).
106. R. M. Sternheimer, *Phys. Rev.* **84**, 244 (1951).
107. R. M. Sternheimer, *Phys. Rev.* **95**, 736 (1954).
108. R. M. Sternheimer, *Phys. Rev.* **105**, 158 (1957).
108a. R. M. Sternheimer, *Phys. Rev. A* **3**, 837 (1971).
109. R. M. Sternheimer, *Phys. Rev.* **146**, 140 (1966).
110. R. M. Sternheimer, *Phys. Rev.* **164**, 10 (1967).
111. R. Ingalls, *Phys. Rev.* **128**, 1155 (1962).
112. A. J. Freeman and R. E. Watson, *Phys. Rev.* **131**, 2566 (1963).
113. A. J. Freeman and R. E. Watson, *Phys. Rev.* **132**, 706 (1963).
114. M. N. Ghatikar, A. K. Raychaudhure, and D. K. Ray, *Proc. Phys. Soc.* **86**, 1239 (1965).

115. G. Heinzelmann, G. Zu Putlitz, and A. Schenck, *Phys. Lett.* **21**, 162 (1966); U. Knohl, G. Zu Putlitz, and A. Schenck, *Z. Phys.* **208**, 364 (1968); G. Zu Putlitz and K. V. Keckataramu, *Z. Phys.* **209**, 470 (1968); H. Buka, G. Zu Putlitz, and R. Rabold, *Z. Phys.* **213**, 101 (1968).
116. K. Murakawa and T. Kamei, *Phys. Rev.* **105**, 671 (1957); K. Murakawa, *Phys. Rev.* **110**, 393 (1958); *J. Phys. Soc. Japan* **16**, 2533 (1961); **17**, 891 (1962).
117. R. E. Watson, *Phys. Rev.* **119**, 170 (1960); A. P. Jucys in *Advances in Chemical Physics*, Vol. XIV, R. Lefebvre and C. Moser, eds. (Interscience Publishers, New York, 1969), p. 191; P. -O. Lowdin, *ibid.*, p. 238.
118. O. Sinanoglu in *Advances in Chemical Physics*, Vol. XIV, R. Lefebvre and C. Moser, eds. (Interscience Publishers, New York, 1969), p. 237.
119. R. K. Nesbet in *Advances in Chemical Physics*, Vol. XIV, R. Lefebvre and C. Moser, eds. (Interscience Publishers, New York, 1969), p. 1; *Phys. Rev. A* **2**, 661 (1970).
120. H. P. Kelly, in *Advances in Chemical Physics*, Vol. XIV, R. Lefebvre and C. Moser, eds. (Interscience Publishers, New York, 1969), p. 129.
121. P. G. H. Sandars in *Advances in Chemical Physics*, Vol. XIV, R. Lefebvre and C. Moser, eds. (Interscience Publishers, New York, 1969), p. 365.
122. K. A. Brueckner, *Phys. Rev.* **97**, 1353 (1955); **100**, 36 (1955); J. Goldstone, *Proc. Roy. Soc. (London)* **A239**, 267 (1957).
123. G. C. Wick, *Phys. Rev.* **80**, 268 (1950).
124. E. S. Chang, R. T. Pu, and T. P. Das, *Phys. Rev.* **174**, 16 (1968).
125. E. S. Chang, R. T. Pu, and T. P. Das, *Phys. Rev.* **174**, 1 (1968).
126. N. C. Dutta, C. Matsubara, R. T. Pu, and T. P. Das, *Phys. Rev. Lett.* **21**, 1139 (1968).
126a. T. Lee, N. C. Dutta, and T. P. Das, *Phys. Rev. A* **1**, 995 (1970).
127. J. D. Lyons, R. T. Pu, and T. P. Das, *Phys. Rev.* **178**, 103 (1969).
127a. R. E. Brown, S. Larsson, and V. H. Smith, Jr., *Phys. Rev. A* **2**, 593 (1970).
128. N. C. Dutta, C. Matsubara, R. T. Pu, and T. P. Das, *Phys. Rev.* **177**, 33 (1969).
129. B. R. Judd and I. Lindgren, *Phys. Rev.* **122**, 1802 (1961); A. Abragam and J. H. Van Vleck, *Phys. Rev.* **92**, 1448 (1953); H. Margenau, *Phys. Rev.* **57**, 383 (1940); G. Breit, *Nature* **122**, 649 (1928).
130. G. Breit and I. I. Rabi, *Phys. Rev.* **38**, 2082 (1931); S. Millman, *Phys. Rev.* **55**, 628 (1939).
131. V. J. Ehlers and H. A. Shugart, *Phys. Rev.* **127**, 529 (1962).
132. W. E. Lamb, *Phys. Rev.* **60**, 817 (1941).
133. W. C. Dickenson, *Phys. Rev.* **80**, 563 (1950).
134. H. Kopfermann, *Nuclear Moments* (Academic Press, Inc., New York, 1958).
135. R. A. Bonham and T. G. Strand, *J. Chem. Phys.* **40**, 3447 (1964).
136. G. Malli and S. Fraga, *Theor. Chem. Acta* **5**, 275 (1966).
137. J. R. P. Angel and P. G. H. Sandars, *Proc. Roy. Soc.* **A305**, 125 (1968).
138. A. Khadjavi, A. Lurio, and W. Happer, *Phys. Rev.* **167**, 128 (1968).
139. J. D. Feichtner, M. E. Hoover, and M. Mizushima, *Phys. Rev.* **137**, A702 (1965).
140. P. G. H. Sandars, *Proc. Phys. Soc.* **92**, 857 (1967).
141. A. Dalgarno, *Advan. Physics* **11**, 281 (1963).
142. N. J. Martin, P. G. H. Sandars, and G. K. Woodgate, *Proc. Roy. Soc.* **A305**, 139 (1968).

References

143. R. Marrus, D. McColm, and J. Yellin, *Phys. Rev.* **147**, 55 (1966). R. Marrus and J. Yellin, *Phys. Rev.* **177**, 127 (1969).
144. J. R. P. Angel, P. G. H. Sandars, and G. K. Woodgate, *J. Chem. Phys.* **47**, 1552 (1967).
145. H. M. Foley, R. M. Sternheimer, and D. Tyco, *Phys. Rev.* **93**, 734 (1954); R. M. Sternheimer and H. M. Foley, *Phys. Rev.* **102**, 731 (1956); R. M. Sternheimer, *Phys. Rev.* **130**, 1423 (1963); **132**, 1637 (1963); **146**, 140 (1966); **159**, 266 (1967), *Phys. Rev.* A **1**, 321 (1970).
146. R. E. Watson and A. J. Freeman, *Phys. Rev.* **131**, 250 (1963); **135**, A1209 (1964).
147. J. Lahiri and A. Mukherji, *Phys. Rev.* **141**, 428 (1966); J. Lahiri, *Phys. Rev.* **153**, 386 (1967).
148. A. Dalgarno and J. M. McNamee, *J. Chem. Phys.* **35**, 1517 (1961); A. Dalgarno and H. A. J. McIntyre, *Proc. Phys. Soc.* **85**, 47 (1965).
149. P. W. Langhoff and R. P. Hurst, *Phys. Rev.* **139**, A1415 (1965).
150. H. P. Kelly and H. S. Taylor, *J. Chem. Phys.* **40**, 1478 (1964); H. P. Kelly, *Phys. Rev.* **136**, B896 (1964).
150a. R. P. Gupta, B. K. Rao, and S. K. Sen, *Phys. Rev.* A **3**, 545 (1971).
151. H. H. Stroke, R. J. Blin-Stoyle, and V. Jaccarino, *Phys. Rev.* **123**, 1326 (1961).
152. H. H. Stroke, in *Atomic Physics*, B. Bederson, V. W. Cohen, and F. M. J. Pichanick, eds. (Plenum Press, New York, 1969).
153. H. M. Foley, in *Atomic Physics*, B. Bederson, V. W. Cohen, and F. M. J. Pichanick, eds. (Plenum Press, New York, 1969).
154. G. H. Fuller and V. W. Cohen, Oak Ridge National Laboratory Report ORNL-4591 (1970).
155. E. Rosenthal and G. Breit, *Phys. Rev.* **41**, 459 (1932).
156. A. Bohr and V. F. Weisskopf, *Phys. Rev.* **77**, 94 (1950).
157. M. F. Crawford and A. L. Schawlow, *Phys. Rev.* **76**, 1310 (1949).
158. For corrections to both Crawford and Schawlow[157] and Bohr and Weisskopf,[156] see N. J. Ionesco-Pallas, *Phys. Rev.* **117**, 505 (1960).
159. A. S. Reiner, *Nucl. Phys.* **5**, 544 (1958).
160. J. E. Eisenger and V. Jaccarino, *Rev. Mod. Phys.* **30**, 528 (1958).
161. E. S. Ensberg, *Bull. Amer. Phys. Soc.* **7**, 534 (1962); Phys. Rev. **153**, 36 (1967).
162. E. Lipworth and P. G. H. Sandars, *Phys. Rev. Lett.* **13**, 716, 718 (1964); J. P. Carrico, E. Lipworth, P. G. H. Sandars, T. S. Stein, and M. C. Weisskopf, *Phys. Rev.* **174**, 125 (1968).
163. J. R. P. Angel, P. G. H. Sandars, and M. H. Tinker, *Phys. Lett.* **25A**, 160 (1967).
164. T. S. Stein, J. P. Carrico, E. Lipworth, and M. C. Weisskopf, *Phys. Rev.* **186**, 39 (1969).
165. C. O. Thornburg, Jr., and J. G. King, *Bull. Amer. Phys. Soc.* **11**, 329 (1966).
166. P. G. H. Sandars, *J. Phys.* **B1**, 499, 511 (1968). For related work, see also L. I. Schiff, *Phys. Rev.* **132**, 2194 (1963); M. Sachs and S. Schwebel, *Ann. Phys. (N.Y.)* **6**, 244 (1959); **8**, 475 (1959); G. Feinberg, *Phys. Rev.* **112**, 1637 (1958).
167. E. Y. Wong, *J. Chem. Phys.* **50**, 4120(L) (1969).
168. E. E. Salpeter, *Phys. Rev.* **112**, 1642 (1958).
169. P. G. H. Sandars, *Phys. Lett.* **14**, 194 (1965); **22**, 290 (1966).

References 199

170. C. Schwartz, *Ann. Phys. (N.Y.)* **6**, 156 (1959).
171. R. D. Haun, Jr. and J. R. Zacharias, *Phys. Rev.* **107**, 107 (1957).
172. E. N. Fortson, D. Kleppner, and N. F. Ramsey, *Phys. Rev. Lett.* **13**, 22 (1964).
173. J. P. Carrico, A. Adler, M. R. Baker, S. Legowski, E. Lipworth, P. G. H. Sandars, T. S. Stein, and M. C. Weisskopf, *Phys. Rev.* **170**, 64 (1968).
174. J. B. Mann, Reports LA-3690 and LA 3691, Los Alamos Scientific Laboratory (1967).
175. C. Schwartz, *Phys. Rev.* **105**, 173 (1957).
176. C. W. Nielson and G. F. Koster, *Spectroscopic Coefficients for p^n, d^n, and f^n Configurations* (MIT Press, Cambridge, Mass., 1963).
177. B. R. Judd, *J. Math. Phys.* **3**, 557 (1962).
178. A. Jucys, J. Vizbaraite, and A. Bandzaitis, *Lietuvos Fizikos Rinkinys* **II**, 123 (1962).
179. L. Armstrong, Jr., *Phys. Rev.* **166**, 63 (1968).
180. W. Albertson, *Phys. Rev.* **52**, 644 (1937).
181. F. M. J. Pichanick and G. K. Woodgate, *Proc. Roy. Soc. (London)* **A263**, 89 (1961).
182. J. G. Conway and B. G. Wybourne, *Phys. Rev.* **130**, 2325 (1963).
183. R. G. H. Robertson, J. C. Waddington, and R. G. Summers-Gill, *Can. J. Phys.* **46**, 2499 (1968).
184. H. A. Bethe and E. E. Salpeter, *Quantum Mechanics of One and Two-Electron Atoms* (Springer-Verlag, Berlin, 1957).

Author Index

Page numbers on which complete references appear are *italicized*.

Abragam, A., 2, 109, 110, 124, *194, 196, 197*
Adler, A., 158, 159, *199*
Akhiezer, A. I., 21, 22, *194*
Albertson, W., 179, *199*
Angel, J. R. P., 132, 133, 134, 135, 150, *197, 198*
Armstrong, L. Jr., 34, 45, 49, 50, *51, 52,* 75, 93, 95, 109, 175, *195, 196, 199*

Back, E., 1, *193*
Bagus, P. 107, 108, 110, *195, 196*
Baker, M. R., 158, 159, *199*
Bandzaitis, A., 173, *199*
Bauche, J., 74, 106, 109, *195*
Bauche-Arnoult, C., 110, *196*
Beck, J., 2, 45, 99, *193*
Bederson, B., 138, *198*
Berestetskii, V. B., 21, 22, *194*
Bernheim, R. A., 2, *193*
Bessis, N., 107, 108, 110, 114, 119, *195, 196*
Bethe, H. A., 189, 191, *199*
Biedenharn, L. C., 13, *194*
Bivens, R., 9, 12, *194*
Bleaney, M. B., 110, *196*
Blin–Stoyle, R. J., 138, 144, *198*
Bloch, F., 2, *194*
Blume, M., 93, *195*
Bohr, A., 7, 138, 141, 144, 145, *194, 198*
Bonham, R. A., 130, *197*
Bordarier, Y., 106, 107, 145, *195*
Breit, G., 1, 21, 124, 125, 138, 144, 145, 191, *193, 197, 198*
Brink, D. M., 8, 10, 13, 14, 58, 59, 60, 77, *194*
Brossel, J., 2, *193*
Brown, R. E., 114, 116, *197*

Brueckner, K. A., 116, 117, *197*
Buka, H., 116, *197*
Burke, E. A., 108, *196*

Carrico, J. P., 150, 156, 158, 159, *198, 199*
Cartan, E., 32, *195*
Casimir, H. B. G., 1, 2, 7, 161, 163, *193*
Chang, E. S., 116, 119, 120, *197*
Childs, W. J., 109, 110, *196*
Cohen, V. W., 138, *198*
Condon, E. U., 2, 9, 21, 49, 74, 83, *194*
Conway, J. G., 179, *199*
Crawford, M. F., 144, 145, *198*
Cromer, D. T., 109, 110, *196*
Crosswhite, H. M., *52*
Crosswhite, Hannah, *52*

Dalgarno, A., 134, 136, *197, 198*
Das, T. P., 108, 116, 119, 120, 121, *196, 197*
DeBroglie, 1, *193*
Desclaux, J. P., 108, *196*
de-Shalit, A., 7, 71, *194*
Dickenson, W. C., 130, *197*
Dirac, P. A. M., 22, *194*
Dutta, N. C., 116, 119, 120, *197*

Edmonds, A. R., 4, 8, 9, 12, *194*
Ehlers, V. J., 126, *197*
Eisenger, J. E., 144, *198*
Elliot, J. P., 13, *194*
Ensberg, E. S., 150, *198*
Evenson, K. M., 110, *196*

Fabry, C., 1, *193*
Feichtner, J. D., 132, 133, 134, 157, *197*
Feinberg, G., 150, *198*
Feneuille, S., 45, 46, 49, 93, *195*

Fermi, E., 1, *193*
Feynman, R. P., 18, *194*
Foley, H. M., 135, 136, 138, *198*
Fortson, E. N., 158, *199*
Fraga, S., 130, *197*
Frankel, R. B., 2, 107, 108, 109, 110, *194, 196*
Frauenfelder, H., 2, *194*
Freeman, A. J., 2, 107, 108, 109, 110, 114, 115, 119, 136, *194, 195, 196, 198*
Fuller, G. H., 138, *198*

Gaspari, G. D., 108, *196*
Gehrcke, E., 1, *193*
Ghatiker, M. N., 115, *196*
Goddard, W. A. III, 108, *196*
Goldstone, J., 116, 117, *197*
Goodman, L. S., 110, *196*
Goren, S. G., 2, *194*
Goudsmit, S., 1, *193*
Gupta, R. P., 136, *198*

Hamermesh, M., 30, *195*
Hansen, W. W., 2, *194*
Happer, W., 132, 133, 134, *197*
Harris, F. E., 107, 108, 110, 114, *195*
Harvey, J. S. M., 110, *196*
Haun, R. D. Jr., 158, *199*
Heinzelmann, G., 116, *197*
Herman, F., 106, *195*
Hoover, M. E., 132, 133, 134, 157, *197*
Horowitz, J., 109, 110, *196*
Hurst, R. P., 136, *198*

Ingalls, R., 115, *196*
Innes, F. R., 77, *195*
Ionesco-Pallas, N. J., 144, *198*

Jaccarino, V., 138, 144, *198*
Jackson, D. A., 1, *193*
Jackson, J. D., 5, 70, 71, *194*
Javan, A., 2, *193*
Jucys, A. P., 13, 14, 116, 117, 173, *197, 199*
Judd, B. R., 13, 14, 15, 16, 21, 30, 34, 35, 36, 40, 48, *51, 52,* 53, 79, 88, 93, 98, 101, 106, 107, 108, 109, 110, 111, 112, 124, 145, 172, 173, *194, 195, 197, 199*

Kamei, T., 116, *197*
Karreman, G., 2, *193*
Kastler, A., 2, *193*
Keckataramu, K. V., 116, *197*
Kelly, H. P., 110, 114, 116, 117, 118, 119, 120, 136, *196, 197, 198*
Kerwin, J., 108, *196*
Keyston, J., 2, *193*
Khadjavi, A., 132, 133, 134, *197*
King, J. G., 150, *198*
Klapisch, M., 106, 107, 145, *195*
Klemm, R. A., 107, 108, 110, 114, *195*
Kleppner, D., 158, *199*
Knohl, U., 116, *197*
Kopfermann, H., 2, 130, 145, 165, *193, 197*
Koster, G. F., 168, 172, *199*
Kusch, P., 2, *193*

Lahiri, J., 136, *198*
Lamb, W. E., 129, 130, *197*
Langhoff, P. W., 136, *198*
Larsson, S., 114, 116, *197*
Lawson, R. D., 30, *194*
Lee, T., 116, *197*
Lefebvre, R., 116, 117, *197*
Lefebvre-Brion, H., 107, 108, 110, 114, 119, *195, 196*
Legowski, S., 158, 159, *199*
Levinson, I. B., 13, 14, *194*
Lewis, W. B., 109, 110, *196*
Liberman, D. A., 109, 110, *196*
Lie, S., 31, *195*
Lindgren, I., 124, *197*
Lipworth, E., 150, 156, 158, 159, *198, 199*
Liu, B., 107, 108, *195*
Low, W., 2, *193*
Lowdin, P. O., 108, 116, 117, *196, 197*
Lummer, O., 1, *193*
Lurio, A., 132, 133, 134, *197*
Lyons, J. D., 116, 120, 121, *197*

McColm, D., 134, *198*
Macfarlane, M. H., 30, *194*
McIntyre, H. A. J., 136, *198*
McNamee, J. M., 136, *198*
Malli, G., 130, *197*
Mann, J. B.,'109, 110, 165, *196, 199*
Margenau, H., 124, *197*
Marrus, R., 75, 134, *195, 198*

Martin, N. J., 134, *197*
Marvin, H. H., 50, *195*
Matsubara, C., 116, 119, 120, *197*
Matthias, E., 2, *194*
Messiah, A., 58, *195*
Metropolis, N., 9, 12, *194*
Michelson, A., 1, *193*
Millman, S., 2, 125, *193, 197*
Mizushima, M., 132, 133, 134, 157, *197*
Moser, C. M., 107, 108, 110, 114, 116, 117, 119, *195, 196, 197*
Mossbauer, R. L., 2, *194*
Mottelson, B. R., 7, *194*
Mukherji, A., 136, *198*
Murakawa, K., 116, *197*

Nesbet, R. K., 107, 108, 110, 114, 116, 117, 118, 119, *196, 197*
Nielson, C. W., 168, 172, *199*
Nutter, P. B., 47, *195*

Packard, M., 2, *194*
Pauli, W., 1, 27, *193, 194*
Perot, A., 1, *193*
Picart, J., 108, *196*
Pichanick, F. M. J., 138, 179, *198, 199*
Pound, R. V., 2, *194*
Preston, M. A., 7, *194*
Pryce, M. H. L., 109, 110, *196*
Pu, R. T., 116, 119, 120, 121, *197*
Purcell, E. M., 2, *194*

Rabi, I., 2, 125, *193, 197*
Rabold, R., 116, *197*
Racah, G., 1, 2, 11, 14, 15, 30, 32, 35, 52, 53, 86, 161, 174, *193, 194*
Radford, H. E., 110, *196*
Rado, G. T., 108, 109, 110, *196*
Rajnak, K., 54, 55, 93, *195*
Ramsey, N. F., 2, 6, 158, *193, 199*
Rao, B. K., 136, *198*
Ray, D. K., 115, *196*
Raychaudhure, A. K., 115, *196*
Reiner, A. S., 144, *198*
Robertson, R. G. H., 179, 183, 185, 186, *199*
Rosen, A., 110, 179, 186, 187, *196*

Rosenthal, E., 138, 144, 145, *198*
Rotenberg, M., 9, 12, *194*

Sachs, M., 150, *198*
Salpeter, E. E., 150, 152, 153, 189, 191, *198, 199*
Sanders, P. G. H., 2, 45, 99, 116, 117, 132, 133, 134, 135, 150, 151, 153, 155, 157, 158, 159, *193, 197, 198, 199*
Satchler, G. R., 8, 10, 13, 14, 58, 59, 60, 77, *194*
Schaefer, H. F. III, 107, 108, 110, 114, *195*
Schawlow, A. L., 144, 145, *198*
Scheffers, G., 31, *195*
Schenck, A., 116, *197*
Schiff, L. I., 150, *198*
Schlichter, C. P., 2, *194*
Schlossberg, H. R., 2, *193*
Schmidt, T., 1, *193*
Schrodinger, E., 1, *193*
Schuler, H., 1, *193*
Schwartz, C., 2, 71, 157, 163, 165, *193, 199*
Schwebel, S., 150, *198*
Sen, S. K., 136, *198*
Shortley, G. H., 2, 9, 21, 49, 74, 83, *194*
Shugart, H. A., 126, *197*
Shyu, W. M., 108, *196*
Sinanoglu, O., 116, 117, *197*
Skillman, S., 106, *195*
Slater, J. C., 2, 21, 49, 74, 83, 194
Smith, P. R., 52
Smith, V. H., 114, 116, *197*
Stein, T. S., 150, 156, 158, 159, *198, 199*
Sternheimer, R. M., 108, 112, 114, 115, 116, 135, 136, *196, 198*
Strand, T. G., 130, *197*
Stroke, H. H., 138, 144, *198*
Suhl, H., 108, 109, 110, *196*
Summers-Gill, R. G., 179, 183, 185, 186, *199*

Talmi, I., 7, 71, *194*
Taylor, H. S., 136, *198*
Thornburg, C. O. Jr., 150, *198*
Tinker, M. H., 150, *198*
Torrey, H., 2, *194*
Trees, R. E., 2, *193*
Tyco, D., 135, 136, *198*

Ufford, C. W., 77, *195*

Vanagas, V. V., 13, 14, *194*
Van der Waerden, B. L., 32, *195*
Van Vleck, J. H., 124, *197*
Vizbaraite, J., 173, *199*

Waddington, J. C., 179, 183, 185, 186, *199*
Wadzinski, H. T., *52*
Watson, R. E., 93, 107, 108, 109, 110, 114, 115, 116, 117, 119, 136, *195, 196, 197, 198*
Weisskopf, M. C., 150, 156, *198, 199*
Weisskopf, V. F., 138, 141, 144, 145, 158, 159, *198*
Wertheim, G. K., 2, *194*
Weyl, H., 32, *195*
Wick, G. C., 116, 118, *197*
Wigner, E. P., 30, *195*
Wolter, H., 112, 114, 115, *196*
Wong, E. Y., 150, 151, *198*
Woodgate, G. K., 110, 134, 135, 146, 179, 182, 183, 185, *196, 197, 198, 199*
Wooten, J. K. Jr., 9, 12, *194*
Wybourne, B. G., 21, *52,* 54, 55, 93, 101, 103, 111, 177, 179, *194, 195, 199*

Yellin, J., 134, *198*
Yutsis, A. P., *see* Jucys, A. P.,

Zacharias, J. R., 2, 158, *193, 199*
Zavoiski, E. J., 2, *194*
Zu Putlitz, G., 116, *197*

Subject Index

A($l_1,l_2; l_3,l_4$), definition, 92
Algebra, Lie, 31
Angular correlation, 2
Angular momenta, coupling of four, 13
  coupling of three, 10–11, 12
  coupling of two, 9
  see also Selection rules for R (3)
Angular momentum, definition, 8, 28, 29
  spectroscopic notation for, 6
Anomaly, hyperfine, *see* Hyperfine anomalies
Anticommutation relations for creation and annihilation operators, 17, 28, 29, 30
Atomic beam, 2
Atomic structure interactions, comparison of relativistic and nonrelativistic forms, 53–55
  group properties of, 51–52
  nonrelativistic form of, 49–50
  relativistic form of, 21, 22

Biedenharn-Elliott sum rule, 13, 146, 148
Bohr-Weisskopf effect, 99, 138, 141–145
Breakdown of J, effects of the, 72, 174
  in Sm, 179–186
  produced by, an external magnetic field, 147–149, 179–183, 184–186
    the hyperfine interaction, 145–148, 174, 181–184
Breit interaction, 21
  group properties of, 50, 51–52
  nonrelativistic limits of, 49–50
  second quantized form of, 43, 44
  in tensor notation, 41–42, 43, 44
Breit-Rabi diagram, 126–127, 128
Breit-Rosenthal effect, 138–141, 142–145
Brueckner-Goldstone linked cluster expansion, 117–121

$C^k(l_1,l_2; l_3,l_4)$, definition, 88, 91, 96
Casimir factors, 161–165
Center of gravity of hyperfine levels, 72–73
Central field, nonrelativistic, 79–84
  relationship to the Hartree-Fock potential, 83
  relativistic, 22, 40, 84–86
Central field approximation, 22, 86–87
  effect on hyperfine structure calculations of, 81, 87, 137–138
  nonrelativistic perturbations in the, 49–52, 81–84
  relativistic perturbations in the, 22, 40–45, 85–86
Clebsch-Gordan coefficient, 9, 49
  generalized, 48–49
Coefficients of fractional parentage, 168
Commutation relations, for angular momenta, 8
  for creation and annihilation operators, 28, 32
  for double tensors, 33
  for the hyperfine hamiltonian, 63, 124
  for infinitesimal operators, 31, 32
  for operators which form a group representation, 46
  for spherical tensors, 14
Contact interaction approximate expression for, 97, 175
  classical, 6–7, 74
  comparison of relativistic and classical forms, 75
  core polarization contribution to, 97–100, 103–104
  correlation contribution to, 119–121
  in Eu, 106–107
  in P, Li, N, and O, 119–120

205

in $3d^n$ atoms and ions, 109–110
in $4d^n$, $5d^n$, $4f^n$ and $5f^n$ atoms and ions, 110
in l^ns configurations, 177
relativistic form of, 65–66, 68, 175
Core polarization, definition, 86, 96
  in Eu, 106–107
  generalized, 118
  Hartree-Fock treatment of, 107–110, 119
  in O, 106, 110, 119–120
  in P, Li, and N, 119–120
  in Sm, 187
  in $3d^n$ atoms and ions, 109–110
  in $4d^n$, $5d^n$, $4f^n$ and $5f^n$ atoms and ions, 110
  importance of continuum in, 106
  perturbation treatment of, 87, 96–107
  radial and angular excitations in, 106, 108, 119
  similarity of relativistic effects to, 104–105
Correlation, definition, 116, 118
  effects, in hyperfine structure of, 117, 118, 119–121
    in the hyperfine structures of O, P, Li, and N, 119–121
  use of, Hartree-Fock techniques to calculate, 117
    many-body techniques to calculate, 108–109, 117–119
Coulomb interaction between electrons, nonrelativistic form, 49, 86, 96
  perturbations involving, 81–82, 83–84, 85, 86–89, 90–91, 97–99, 100, 101–103, 104, 111
  relativistic tensor form, 41, 43–44, 89, 91
  second quantized form, 43–44
Creation and annihilation operators, as group generators, 32–33
  coupled, 29–30
  for nonrelativistic electrons, 16–17
  for relativistic (sl) electrons, 28–29
  for relativistic (slj) electrons, 27–28
  as representations of groups, 46
  tensorial properties of, 28, 29

Diamagnetic shielding, 129–130
Differential Polarizability, 156–159
Dirac equation, 2, 22, 23–27
  for an electron with an electric dipole moment, 152

gauge transformations and, 59
nonrelativistic limit of, 26, 189–191
second order, 189
vector potentials in, 58, 123
Dot product of two spherical tensors, 3, 15
Double tensors, as bases for representations of groups, 46–47, 168–169
  commutation relations for, 33
  as group generators, 33, 45, 171
  nonrelativistic, 15, 45, 168
  relativistic, 30

Effective operators, 20, 67–68, 89
  for breakdown of J, 147, 148, 149
  for core polarization, 103–104
  for differential polarizability, 157–159
  for interactions with an electric field, 131, 133, 157–159
  for perturbations involving one-electron operators, 88–89, 90–92, 94–96
  for quadrupole shielding, 111
  for relativistic atomic structure operators, 45
  for relativistic hyperfine interactions, 2, 68, 69–71
  for the spin-orbit interaction, 93–94
Electric dipole moment, enhancement factor R, 155–156
  of atom, 150–156
  of electron, 152
  perturbation operator for an electron having, 152–153
  vanishing for a nonrelativistic atom of, 154
Electric quadrupole interaction, 1–2
  classical derivation of, 3–5
  differential polarizability and, 158
  effective operators for, 68
  in the half filled shell, 174
  nonrelativistic form of, 75–76
  relativistic and nonrelativistic forms compared, 76
  relativistic form of, 66
  shielding of, see Quadrupole shielding
  with an external electric field, 135–136
Electric quadrupole interaction constant b, approximate one electron, 165–166
  in the configuration $l^n l'm$, 177–178
  effect of breakdown of J on, 147, 174
  nonrelativistic form of, 76, 134
  one-electron, 160
  relativistic form of, 70–71

Subject Index

Electromagnetic fields of the nucleus,
    scalar potential, 58
    vector potential, 58–62
External electric field, hamiltonian for, 131, 133, 135, 157
External magnetic field, hamiltonian for, 123, 124

$F^{(k)}$, definition, 63
Feynman graphs, for differential polarizability, 157
    for perturbations to the operator $T^{(SL)J}$, 93
    rules for drawing, 18–19
    for second-order Stark effect, 132
Filled shell, 18–19
Fine structure, relation of hyperfine structure to, 1, 2, 57
    see also Atomic structure
Fractional parentage, coefficients of, 168

Gauge transformations, 58–59, 61–62
Group, basis for, 34
    continuous, 31
    creation and annihilation operators as generators of, 32–33
    definition of, 30–31
    double tensors as generators of, 33
    electronic states as a basis of a representation of, 35, 36–39, 45, 48, 54
    generators of, 31, 32–34
    invariant, 32
    Lie, 31
    operators as representations of, 45–48, 50–52
    representations of, 34–35
    simple, 32
    subgroups of, 31, 33, 34
    see also Selection rules

Half filled shell, 76, 174–175
Hartree-Fock, configuration interaction method, 107–108
    core polarization and, 107–110, 119
    correlation studies and, 117, 118, 119
    diamagnetic shielding and, 130
    one-electron excitations and, 83–84, 116
    potential, 83
    quadrupole shielding and, 114
    relativistic, 161, 165, 167, 187
    restricted, 84
    spin and orbitally polarized, 108, 109, 119, 136
Hyperfine anomalies, approximate expression for, 144, 145
    definition, 138
    in Eu, 145
    nuclear magnetism contribution to, 99, 141–145
    nuclear volume contribution to, 138–141, 142–145
    in Sm, 185, 187
Hyperfine hamiltonian, classical derivation, 3–7
    core polarization and relativistic effects compared, 104–105
    in an electric field, 133–134
    including complete vector potential of nucleus, 141
    including effect of breakdown of J, 147, 148, 149
    including effect of core polarization, 81, 103–104
    in a magnetic field, 124–125
    nonrelativistic, 73–78
    relativistic, 1, 62–63, 64–65, 68
    see also Electric quadrupole interaction; Magnetic dipole interaction; Magnetic octupole interaction
Hyperfine structure, definition, 2, 57
    relation to fine structure, 1, 2, 55–56, 57

Interval rule for hyperfine structure, 72
Isotope shift, 2

Kronecker product, 47
    of R(3), 68–69

Lande g factor, 122–123

$M^{(k)}$, definition, 65
Magnetic dipole interaction, 1, 7
    classical derivation, 5–7
    contribution to differential polarizability, 158
    effective operators for, 68
    effect of core polarization on, 103–104, 109–110
    in a half-filled shell, 175
    nonrelativistic form of, 74

relativistic and core polarization effects compared, 104–105
relativistic form of, 65–66
relativistic and nonrelativistic forms compared, 75
see also Contact interaction
Magnetic dipole interaction constant A, in the configuration $l^n l'm$, 176–177
effects of core polarization on, 96–107, 109–110, 119–121
effects of correlation on, 119–121
nonrelativistic form of, 74–75
one electron, 160, 165–166
relativistic form of, 69–70, 137
see also Hyperfine anomalies
Magnetic dipole moment of electron, 6
Magnetic octupole interaction, 2
contribution to differential polarizability, 158
effective operators for, 68
nonrelativistic form of, 77
relativistic form of, 66–67
relativistic and nonrelativistic forms compared, 78
Magnetic octupole interaction constant c, in the configuration $l^n l'm$, 177–178
effects of breakdown of J on, 147
one electron, 160–161, 165–166
nonrelativistic form of, 78
relativistic form of, 71
Mossbauer effect, 2

$N^{(k)}$, definition, 63
Nine-j symbols, 13–14
Nuclear electric quadrupole moment Q, 1, 2, 71, 135
definition, 71
Nuclear magnetic dipole moment μ_I, 1, 122
correction for diamagnetic shielding, 129–130
definition, 5, 70
interaction with external magnetic field, 122
Nuclear magnetic octupole moment Ω, 2, 71
definition, 71
Nuclear magnetic resonance, 2
Nuclear models, 7

Optical pumping, 2
Optical spectroscopy, 1
Orbit-orbit interaction in l^n, 49–50, 51–52

Paramagnetic resonance, 2
Parity, of hyperfine operators, 3, 63
operator in four-component theory, 24
selection rules bases on, 3, 60, 63, 130
violation in atomic structure of, 150
Passive shells, definition, 20
excitation in the central field model and, 20, 82–83, 85, 86
Pauli principle, 18, 27, 29
Polarizability, differential, 156–159
tensor and scalar, 133–134, 157

$Q^{(k)}$, definition, 64
Quadrupole shielding, 87
angular and radial excitation in, 112, 113–114, 136
in Cs, Cu, and Rb, 116
excitations to the continuum in, 112, 134
Hartree-Fock method of calculation, 114, 136
in Li, 119–120
in O, 112, 119–120
in $2p^n$ atoms, 112
perturbation calculation, 110–112
relative effects of direct and exchange interactions, 114
shielding factor R, 111, 114–115
shielding factor γ_∞, 135–136
Sternheimer method of calculating, 87, 112–114, 136
Quasispin, definition, 33
relation to seniority, 36
selection rules based on, 168, 174
used to evaluate matrix elements, 173–174
used to label atomic states, 36–39

$R^k(n_1 l_1, N_2 l_2; n_3 l_3, n_4 l_4)$, definition, 80
Racah algebra, Racah technique, 2, 53
Radial integrals, relativistic correction factors to, 163–165
parametrization of, 53, 54–55, 167, 179, 185, 186
Reduced matrix element, for coupled operators, 15, 16

definition, 15
  of nonrelativistic one-body operators, 15, 16, 45, 168, 171, 172, 173, 174
  of relativistic one-body operators, 30, 43, 44
Relativistic correction factors, 161–165
Representations, group, 34–35, 36–39, 45–48, 50–52
Rosenthal-Breit effect, 138–141, 142–145

Scalar polarizability, 133–134, 157
Scalar product of spherical tensors, 3, 15
Second quantization formalism, 16–18
  applied to perturbation theory, 81–83, 88–89, 99
  Feynman graphs and, 18–20
  operators in, 17, 18, 19, 30, 40, 43, 64, 89
Selection rules, based on parity, 60, 63, 130
  for G_2, 169–171
  for R(3), 9, 12, 13, 48, 63, 168
  for R(5), 168–170, 171
  for R(7), 168–169
  for Sp(4l+2), quasispin, seniority, 168, 174
  Kronecker products and, 47–48
Seniority, 36; *see also* Quasispin
Shell, electronic, 18, 29, 35, 105, 108
Six-j symbols, 11–13
Spherical tensors, Breit interaction expressed in terms of, 41–42
  coupling of, 14
  creation and annihilation operators as, 28, 29
  definition, 14
  hyperfine interaction expressed in terms of, 63, 68, 69–71, 72
  reduced matrix elements and, 15
  as representations of R(3), 46
  scalar product of two, 3, 15
Spin-orbit interaction, nonrelativistic form of, 49, 50, 51–52, 92–93, 95–96
  perturbations involving, 86, 89, 90–91, 91–92, 93–94, 95–96, 100, 101, 102, 103, 104, 111–112
  relativistic form of, 89, 92
  two-body nature of, 93–94, 95–96
Spin-other-orbit interaction in l^n, 50, 51–52
Spin-spin contact interaction in l^n, 50, 51–52
Spin-spin interaction in l^n, 50, 51–52
Stark effect, 130–134, 156–159
Sternheimer shielding, 87, 112–114, 136

Tensor polarizability, 133–134, 157
Three-j symbols, 9–11
Time reversal, violation in an atom of, 151–152
Triangular conditions, 9, 12, 13, 48, 63, 168

Vector spherical harmonics, 59–62

Wigner-Eckart theorem, definition, 48–49
  in quasispin space, 173
  in R(3), 15, 49, 69, 70, 71, 93, 99, 101, 123, 124–125, 151
  reduced matrix elements and, 15

/539.14A736T>C1/